Dell VxRail System Design and Best Practices

A complete guide to VxRail appliance design and best practices

Victor Wu

BIRMINGHAM—MUMBAI

Dell VxRail System Design and Best Practices

Group Product Manager: Mohd Riyan Khan
Publishing Product Manager: Shrilekha Malpani
Senior Content Development Editor: Sayali Pingale
Technical Editor: Shruthi Shetty
Copy Editor: Safis Editing
Book Project Manager: Neil Dmello
Proofreader: Safis Editing
Indexer: Sejal Dsilva
Production Designer: Ponraj Dhandapani
Marketing Coordinator: Ankita Bhonsle
Senior Marketing Coordinator: Marylou De Mello

First published: December 2022

Production reference:1301122

Published by Packt Publishing Ltd.
Livery Place
35 Livery Street
Birmingham
B3 2PB, UK.

978-1-80461-770-0

www.packt.com

Contributors

About the author

Victor Wu has over 15 years of system infrastructure experience. Currently, he works as a senior solutions architect at BoardWare Information System Limited in Macau.

He is the only qualified person in Macau with a certificate in VMware VCIX-DCV, and was awarded the vExpert certification from 2014 to 2022, Cisco Champion from 2017 to 2022, Veeam Vanguard from 2019 to 2022, and Nutanix Technology Champion in 2021.

His professional qualifications include VCIX-DCV 2022, VMware Certified Master Specialist – HCI 2022, Implementation Engineer – VxRail Appliance, Systems Administrator – VxRail Appliance, Nutanix Certified Professional 5, NetApp HCI Implementation Engineer, and Knowledge Sharing Author from 2018 to 2021.

He is the author of *Implementing VxRail HCI Solutions*, published by Packt Publishing in June 2021.

About the reviewers

Venkata Krishna Mallemarapu is a senior systems integrations advisor with over 13 years of experience in IT. He holds certifications in the fields of hyper-converged infrastructure, virtualization, storage, and networking. Krishna completed his education in the fields of computer science, information technology, electronics, and biomedical engineering.

Krishna is a people-friendly person who loves to entertain. He loves exploring new places, playing ping-pong, volleyball, and tennis, and gardening.

I would like to dedicate this to my parents, Siva Kumari and Rathaiah. Thank you for the support, the unconditional love, and for always being there for me.

To my wife, Sruthi. Thank you for your love, support, and encouragement. I am so blessed to spend the rest of my life with you.

To my friends. Thank you for the birthdays, inside jokes, food, laughs, and all the board and card game nights we've shared.

Pradeep Adapa has worked in IT for more than 11 years, with more than 9 years specifically dedicated to virtualization. He is a blogger with a blog dedicated to hyper-converged systems, especially **VMware Cloud Foundation** (**VCF**) and VxRail. He has a master's in information systems from Marist College, Poughkeepsie, NY. He is currently employed at Solera Inc, full-time, where he works primarily on VMware Cloud Foundation 3.11.x and vRealize products (VRA, vRealize Log Insight, vRealize Operations, VRNI, and VRSLCM). He has also been a vExpert for two consecutive years (2021–2022).

I'd like to thank my wife, Hari, for putting up with me during this journey of reviewing the book and contributing to reviewing the chapters in this book.

Table of Contents

Part 2: Design of the VxRail Appliance 7.x System

3

4

5

6

Design of vSAN 2-Node Cluster on VxRail   135

Part 3: Design of Data Protection for the VxRail System

7

8

9

Design of RecoverPoint for Virtual Machines on VxRail   225

10

Design of VxRail with Veeam Backup   257

Preface

Traditional IT teams are faced with a massive amount of complexity when building, configuring, maintaining, and scaling applications. **Hyper-Converged Infrastructure** (**HCI**) can simplify infrastructure deployment and management. VxRail appliances are developed by Dell EMC and VMware and are the only fully preconfigured and tested HCI appliances powered by VMware vSAN technology. This book contains three sections, getting started with the VxRail appliance 7.x system, the design of the VxRail appliance 7.x system, and the design of the data protection for the VxRail system. You will be given an overview of VxRail's architecture and the benefits of the VxRail 7.x system, and design every core component of a VxRail system, the vSAN storage policies, and its cluster expansion. In the last section, you will learn about the design of the advanced solutions for VxRail, including a stretched cluster on VxRail, VMware **Site Recovery Manager** (**SRM**), Dell EMC RecoverPoint for VMs, and Veeam Backup & Replication.

By the end of this book, you will have got to grips with Dell's hyper-converged VxRail offering, taking your virtualization proficiency to the next level.

Who this book is for

This book is for system architects, system administrators, or consultants involved in planning and designing VxRail HCI. The reader is expected to have equivalent knowledge and administration experience with VMware vSphere 7. x and vCenter Server 7.x.

What this book covers

Chapter 1, Overview of VxRail Appliance 7.x System, provides an overview of the VxRail appliance 7.x system. The architecture of Dell HCI is different from a traditional server and storage.

Chapter 2, Benefits of Hyper-Converged Infrastructure, outlines the benefits of the VxRail appliance 7.x system and Dell EMC VxRail on the 15th-generation PowerEdge portfolio.

Chapter 3, Design of vCenter Server, describes the design of VxRail deployment options – for example, an internal vCenter Server with an external DNS, an internal vCenter Server with an internal DNS, an external vCenter Server with an external DNS, and so on.

Chapter 4, Design of vSAN Storage Policies, explains the design of vSAN storage policies on the VxRail appliance system.

Chapter 5, Design of Cluster Expansion, details the VxRail scale-up and scale-out rules, and how to design cluster expansion.

Chapter 6, Design of vSAN 2-Node Cluster on VxRail, clarifies the best practices for a vSAN 2-Node cluster on VxRail.

Chapter 7, Design of Stretched Cluster on VxRail, gives an overview of a stretched cluster on VxRail and how to design this advanced solution.

Chapter 8, Design of VxRail with SRM, explores the disaster recovery solutions for VxRail. It includes the active-passive solution with VMware SRM.

Chapter 9, Design of RecoverPoint for Virtual Machines on VxRail, describes the **Continuous Data Protection** (**CDP**) solution for VxRail and teaches the reader how to plan and design this advanced solution.

Chapter 10, Design of VxRail with Veeam Backup, demonstrates how the Dell VxRail system is also supported by a third-party backup solution. You will learn how to design VxRail with Veeam Backup & Replication.

To get the most out of this book

Make sure your workstation (laptop) is running on the Windows platform and a web browser is installed onto your laptop. The latest versions of Firefox, Google Chrome, and Microsoft Internet Explorer 11 or above are all supported. You need to run the VxRail software at version 7.0.320 or above.

Software/Hardware covered in the book	OS Requirements
Microsoft Windows platform	Windows 8 or 10
VxRail software	VxRail 7.0.320 or above
VMware SRM and VR	Version 8.3 or above
Dell EMC RecoverPoint for VMs	Version 5.3 or above
Veeam Backup & Replication	Version 12 or above

Download the color images

We also provide a PDF file that has color images of the screenshots/diagrams used in this book. You can download it here: `https://packt.link/8W828`.

Conventions used

There are a number of text conventions used throughout this book.

`Code in text`: Indicates code words in text, database table names, folder names, filenames, file extensions, pathnames, dummy URLs, user input, and Twitter handles. Here is an example: "The preferred site advertises the `192.169.0.0/24` network to the third site, and the secondary site advertises the `192.170.0.0/24` network to the third site."

A block of code is set as follows:

```
Install-WindowsFeature Web-Application-Proxy
-IncludeManagementTools
```

Bold: Indicates a new term, an important word, or words that you see onscreen. For example, words in menus or dialog boxes appear in the text like this. Here is an example: "You can select the following options from the **Failures to tolerate** menu."

> **Tips or important notes**
> Appear like this.

Get in touch

Feedback from our readers is always welcome.

General feedback: If you have questions about any aspect of this book, mention the book title in the subject of your message and email us at `customercare@packtpub.com`.

Errata: Although we have taken every care to ensure the accuracy of our content, mistakes do happen. If you have found a mistake in this book, we would be grateful if you would report this to us. Please visit `www.packtpub.com/support/errata`, selecting your book, clicking on the Errata Submission Form link, and entering the details.

Piracy: If you come across any illegal copies of our works in any form on the Internet, we would be grateful if you would provide us with the location address or website name. Please contact us at `copyright@packt.com` with a link to the material.

If you are interested in becoming an author: If there is a topic that you have expertise in and you are interested in either writing or contributing to a book, please visit `authors.packtpub.com`.

Share Your Thoughts

Once you've read *Dell VxRail System Design and Best Practices*, we'd love to hear your thoughts! Scan the QR code below to go straight to the Amazon review page for this book and share your feedback.

https://packt.link/r/1804617709

Your review is important to us and the tech community and will help us make sure we're delivering excellent quality content.

Download a free PDF copy of this book

Thanks for purchasing this book!

Do you like to read on the go but are unable to carry your print books everywhere?

Is your eBook purchase not compatible with the device of your choice?

Don't worry, now with every Packt book you get a DRM-free PDF version of that book at no cost.

Read anywhere, any place, on any device. Search, copy, and paste code from your favorite technical books directly into your application.

The perks don't stop there, you can get exclusive access to discounts, newsletters, and great free content in your inbox daily

Follow these simple steps to get the benefits:

1. Scan the QR code or visit the link below

https://packt.link/free-ebook/978-1-80461-770-0

2. Submit your proof of purchase
3. That's it! We'll send your free PDF and other benefits to your email directly

Part 1: Getting Started with the VxRail Appliance 7.x System

In this part, the reader will get an overview of the VxRail appliance 7.x system; this includes the architecture, features, and documentation resources and what the benefits of the VxRail Appliance 7.x System are.

This part of the book comprises the following chapters:

- *Chapter 1, Overview of VxRail Appliance 7.x System*
- *Chapter 2, Benefits of Hyper-Converged Infrastructure*

1

Overview of VxRail Appliance 7.x System

In the digital economy, most applications need to provide a 24*7 **Service-Level Agreement** (**SLA**) for every customer. The IT service department provider often faces the problems of how to make applications available at any time, how to handle **Life Cycle Management** (**LCM**), how the system infrastructure can automatically scale up and out, and so on. Most traditional infrastructure architecture has some hardware and software limitations; it cannot fulfill these requirements. You may need to integrate third-party hardware and software to fulfill these requirements. However, this will increase the cost of the total solution. When new software packages or patches are released, they need to upgrade or apply the patch to the existing traditional infrastructure architecture. At this moment, some compatibility problems between the hardware and software may exist, which is why traditional infrastructure architecture is not a good solution.

With the **Hyper-Converged Infrastructure** (**HCI**) platforms available on the market, most technical limitations of traditional infrastructure architecture can be resolved. The HCI platform can simplify most day-one deployment and day-two management activities. Dell VxRail Appliance is an HCI platform developed by Dell EMC and VMware. VxRail Appliance can provide different features, for example, hardware scaling, software package upgrade, centralized management, and LCM. In this chapter, we will discuss the VxRail system; you will get an overview of the VxRail Appliance 7.x platform.

This chapter includes the following main topics:

- What is VxRail Appliance?
- Dell VxRail architecture
- Dell VxRail features
- Dell VxRail management
- Dell VxRail documentation and resources

What is VxRail Appliance?

VxRail Appliance (as seen in *Figure 1.1*) is developed and powered by Dell EMC and VMware. It is an HCI appliance that is exclusively integrated and preconfigured with VMware vSphere and **Virtual SAN (vSAN)**. VxRail platforms are fully integrated with **VMware vCenter Server Appliance (VCSA)** and use a VxRail Manager plugin for vCenter. The VxRail platform provides user-friendly and centralized management; system administrators can perform day-to-day activities using the VxRail Manager plugin for vCenter in a **Hypertext Markup Language revision 5 (HTML5)** interface:

Figure 1.1 – VxRail system on the Dell 15th-generation PowerEdge server

The VxRail platform is powered by Dell EMC PowerEdge servers with Intel Scalable or AMD EPYC processors. VxRail Appliance can be configured with different hardware options, for example, CPU processors with different cores, different sizes of memory, the network connectivity interface, a Fibre Channel **Host Bus Adapter (HBA)**, a **Graphics Processing Unit (GPU)**, Hybrid mode, and All-Flash mode of disks. VxRail Appliance is fully integrated with VMware solutions including VMware Tanzu, VMware NSX, VMware vRealize Suite, and VMware **Site Recovery Manager (SRM)**, and private clouds such as **VMware Cloud Foundation (VCF)**. You can check out the VMware website (`https://www.vmware.com`) if you want to learn about these VMware products in more detail.

VxRail Appliance models are available in different form factors, and they come as one unit per node, two units per node, and four nodes in a two-unit chassis. The VxRail Appliance architecture is designed so the customer can buy and scale out based on their infrastructure requirements. Dell solution architects use VxRail Sizing Tool (`https://vxrailsizing.emc.com`) for VxRail design. VxRail Sizing Tool is an online platform that analyzes workloads and hardware requirements, then provides a VxRail hardware configuration that meets customer requirements. The Dell EMC VxRail family offers six types of platforms, that is, E (entry-level) Series, P (performance-optimized) Series, V (VDI-optimized) Series, D (durable-platform) Series, S (storage-dense) Series, and G (general-purpose) Series.

Now, we will introduce each model of Dell EMC VxRail Appliance.

VxRail E Series has three options, Hybrid, All-Flash, and NVMe. You can choose the All-Flash or NVMe model if you want high performance. For general purposes, you can choose the Hybrid model. Each node is a **one-unit** form factor system that is used for most scenarios. It is based on Dell EMC PowerEdge R650/R6515 server technology. VxRail E Series includes two models, **VxRail E660** and **E665**. VxRail 660 runs on the Dell EMC PowerEdge R650 server, which supports 10 GB, 25 GB, and 100 GB network interfaces, and VxRail 665 runs on the Dell EMC PowerEdge R6515 server, which supports only 10 GB and 25 GB network interfaces.

Figure 1.2 – VxRail E Series on the Dell 15th-generation PowerEdge server

VxRail P Series only has two options, All-Flash or NVMe. Each node is a **two-unit** form factor system that is used for high-performance and data-intensive application scenarios. It is based on Dell EMC PowerEdge R750/R7515/R840 server technology; VxRail P Series includes three models, **VxRail P670**, **P675**, and **P580N**. It also supports 10 GB, 25 GB, and 100 GB network interfaces. VxRail P670 (All-Flash mode only) runs on the Dell EMC PowerEdge R750 server, which supports 10 GB, 25 GB, and 100 GB network interfaces. VxRail P675 (All-Flash and NVMe mode) runs on the Dell EMC PowerEdge R7515 server, which supports 10 GB, 25 GB, and 100 GB network interfaces. VxRail P840N (NVMe mode only) runs on the Dell EMC PowerEdge R580N server, which supports 10 GB, 25 GB, and 100 GB network interfaces.

Figure 1.3 – VxRail P Series on the Dell 15th-generation PowerEdge server

VxRail V Series only has All-Flash mode. Each node is a **two-unit** form factor system that is used for VDI optimized for specialized scenarios. It is based on Dell EMC PowerEdge R750 server technology; VxRail V Series includes only one model, **VxRail V670**. VxRail V670 (All-Flash mode only) supports 10 GB, 25 GB, and 100 GB network interfaces. Only V Series can support GPU cards.

Figure 1.4 – VxRail V Series on the Dell 15th-generation PowerEdge server

VxRail D Series only has two options, Hybrid or All-Flash. Each node is a **one-unit** form factor system that is designed to withstand extreme conditions, for example, intense heat, cold, or humidity. It is based on Dell EMC PowerEdge XR2 server technology; VxRail D Series includes only one model, **VxRail D560**. VxRail D560 only supports 10 GB and 25 GB network interfaces. VxRail D560 is available in MIL-STD and DNV-GL Maritime-certified configurations.

Figure 1.5 – VxRail D Series on the Dell 15th-generation PowerEdge server

VxRail S Series only has a Hybrid option. Each node is a **two-unit** form factor system that is used for demanding applications, for example, big data, Microsoft Exchange, and archive data. It is based on Dell EMC PowerEdge R750 server technology; VxRail S Series includes only one model, **VxRail S670**. VxRail S670 supports 10 GB, 25 GB, and 100 GB network interfaces.

Figure 1.6 – VxRail S Series on the Dell 15th-generation PowerEdge server

VxRail G Series has two options, Hybrid or All-Flash. Each node is a **two-unit** form factor system that is used for general-purpose virtualized workloads. It is based on Dell EMC PowerEdge C6420 server technology; VxRail G Series includes only one model, **VxRail G560**. VxRail G560 supports 10 GB, 25 GB, and 100 GB network interfaces. The VxRail G Series chassis can install four nodes.

Figure 1.7 – VxRail G Series on the Dell 15th-generation PowerEdge server

> **Important Note**
> In the initial deployment, the first three VxRail nodes in a cluster must be identical models. VxRail Hybrid and All-Flash nodes cannot mix in a VxRail cluster.

The following table shows a summary of hardware configurations on each VxRail Series:

VxRail Series	E Series	P Series	V Series	D Series	S Series	G Series
Roles	Entry-level	High performance	VDI scenario	Durable and rugged	Dense storage	Dense compute
Options	Hybrid, All-Flash, NVMe	All-Flash, NVMe	All-Flash	Hybrid, All-Flash	Hybrid	Hybrid, All-Flash
Models	E660, E665	P670, P675, P580N	V670	D560	S670	G560
Unit	1-unit	2-unit	2-unit	1-unit	2-unit	1-unit
CPU	1 or 2 CPUs	1 or 2 CPUs	1 or 2 CPUs	1 or 2 CPUs	1 or 2 CPUs	1 or 2 CPUs
Memory	64 GB to 4,096 GB	64 GB to 6,144 GB	128 GB to 4,096 GB	64 GB to 1,024 GB	64 GB to 4,096 GB	64 GB to 2,048 GB
Disk slots	10	28	24	8	16	6
Disk Groups	2	4	4	2	4	1
Onboard Networking	10 GB SFP+, BASE-T, 25 GB SFP28	10 GB SFP+, BASE-T, 25 GB SFP28	10 GB SFP+, BASE-T, 25 GB SFP28	10 GB BASE-T, 25 GB SFP28	10 GB SFP+, BASE-T, 25 GB SFP28	10 GB SFP+
Additional Network	Supported	Supported	Supported	Supported	Supported	Supported
Fibre Channel	16 GB, 32 GB HBA	16 GB, 32 GB HBA	16 GB, 32 GB HBA	N/A	16 GB, 32 GB HBA	N/A
GPU	Supported	Supported	Supported	Supported	N/A	N/A

Table 1.1 – A summary of hardware configurations on each VxRail Series

The VxRail Appliance 7.x platform supports three types of VxRail nodes, including a VxRail node with vSAN, a VxRail dynamic node, and a VxRail satellite node. We will discuss the details in *Chapter 2, Benefits of Hyper-Converged Infrastructure*.

What is inside VxRail Appliance?

VxRail Appliance is a black-box solution developed by Dell EMC and VMware. VMware vSphere is preinstalled and configured on each VxRail Appliance before shipping to customers. In VxRail Appliance, there are three key elements:

- VMware SDDC technologies
- VxRail HCI system software
- Data protection options

This table shows a summary of the software inside VxRail Appliance:

VMware SDDC Technologies	VxRail HCI System Software	Data Protection Options
Choice of vSAN	VxRail Manager	Dell EMC RecoverPoint for VMs
vCenter Server	SaaS multicluster management	VMware vSphere Replication
vRealize Suite Ready	RESTful APIs	N/A
vSphere Ready	Automation and orchestration services	N/A
Tanzu Basic (optional)	Ecosystem connectors	N/A
VMware Cloud Foundation (optional)	N/A	N/A

Table 1.2 – A summary of the software inside VxRail Appliance

With the preceding table, we know what components make up each key element. Now we will discuss each piece of software.

VMware SDDC technologies

The following is a list of SDDC technologies:

- **VMware vSAN** is **Software-Defined Storage** (**SDS**) that is embedded in the kernel with VMware vSphere. You can choose the edition of VMware vSAN based on your requirements when you buy VxRail Appliance, including Standard, Advanced, Enterprise, and Enterprise Plus. Each VxRail Appliance is enabled by the vSAN feature when you power it on for the first time.

> **Important Note**
> In VxRail 4.7 or above, the VMware vSAN license is not embedded in each VxRail node. The vSAN license is enabled in the evaluation mode on each node when the customer powers it on for the first time. They need to add the vSAN license manually to each node.

- **VMware vCenter Server** is a central management dashboard that is used to configure and manage virtual machines. The VCSA instance license is preinstalled on VxRail Appliance. VCSA is a virtual appliance that is fully integrated with VxRail Appliance using the VxRail Manager plugin; we can manage VxRail appliances via VMware vCenter Server with the VxRail Manager plugin.

> **Important Note**
> The embedded vCenter instance license is bundled on VxRail Appliance. If you deployed the external vCenter Server for VxRail management, the optional vCenter Server instance license is required.

- **VMware vRealize Suite** is a cloud-based management software package that integrates VMware vRealize Cloud Management products, including vRealize Automation, vRealize Operations, vRealize Log Insight, and LCM on-premises. You can enable VMware vRealize products on the VxRail Appliance system at any time.

- **VMware ESXi** is a hypervisor that virtualizes the physical resources (for example, CPU processors, memory, storage, and networking) on the different virtual machines.

> **Important Note**
>
> VMware vSphere licenses are not included in VxRail Appliance. The customer can reuse existing vSphere licenses or order new licenses for each VxRail node.

- **VMware Tanzu** is used for modernizing your applications, which helps customers execute and manage different **Kubernetes** (**K8s**) clusters across the multicloud platform.

- **VMware Cloud Foundation** (**VCF**) is used to build up the **Software-Defined Data Center** (**SDDC**). This platform can deliver a variety of features, including VMware vSAN, VMware NSX, and VMware vRealize features.

> **Important Note**
>
> VMware Tanzu and VCF are optional features on VxRail; both features are not included in VxRail Appliance.

VxRail HCI system software

The following is a list of VxRail HCI system software:

- **VxRail Manager** is predeployed on VxRail Appliance, and it is fully integrated into VCSA with the VxRail Manager plugin. You can execute daily operations via the VxRail Manager plugin; these tasks include system scale-up and scale-out, automatic deployments, LCM, and maintenance tasks.

- **SaaS multicluster management** is a Dell EMC cloud platform. SaaS multicluster management collects telemetry data from each node in the VxRail cluster via the **Secure Remote Services** (**SRS**) gateway and can deliver the proactive system monitoring of VxRail Appliance systems.

- **RESTful APIs** are a bundled feature on VxRail Appliance that can execute management functions.

- **Automation and orchestration services** include VxRail Manager delivering the automated deployment and orchestration workflow in the VxRail Appliance system, for example, system scale-up and scale-out, LCM, and an active-active solution.

- **Ecosystem connectors** are used to integrate the software and hardware for LCM in the VxRail cluster, including automation and orchestrating services.

Data protection options

Dell EMC **RecoverPoint for VMs (RP4VM)** is a **Continuous Data Protection (CDP)** solution from Dell EMC. It can provide protection of the virtual machine with its point-in-time synchronization or asynchronization in a local VxRail cluster or across VxRail clusters between two different locations. You can manage RP4VM directly via VCSA.

> **Important Note**
>
> The RP4VM license includes 5 virtual machine licenses per node (E, P, V, D, and S Series) and 15 virtual machine licenses per chassis for the G Series.

VMware **vSphere Replication (VR)** is a disaster recovery solution at the hypervisor level; it can deliver data protection of the virtual machine with a 5-minute **Recovery Point Objective (RPO)** in a local VxRail cluster or across VxRail clusters between two different locations. You can manage VR directly via VCSA, and it also can work with SRM to deliver the automated failover and failback recovery plan.

> **Important Note**
>
> VMware SRM is the optional license for enabling disaster recovery on VxRail Appliance. You need an SRM license with at least 25 virtual machines per site.

Now that we understand which software components are bundled with VxRail Appliance, let's look at the licensing options.

VxRail licensing

VxRail Appliance is a turnkey solution that is bundled with some software licenses on each VxRail node. It includes the following bundled VMware and Dell EMC software licenses:

- VMware vCenter Server
- VMware vSphere
- VMware vSAN
- VMware vRealize Log Insight
- VMware VR

Dell EMC software includes the following:

- Dell EMC RecoverPoint for Virtual Machines

VxRail Appliance also follows a **Bring-Your-Own** (**BYO**) vSphere license model. You can purchase VxRail Appliance with a new vSphere license or reuse any existing qualified vSphere licenses. The VxRail system supports several VMware vSphere license editions; it includes Enterprise Plus, Standard, and **ROBO** (**Remote Office Branch Office**) editions.

This table shows the difference between the preceding-supported vSphere licenses:

Edition	vSphere Enterprise Plus	vSphere Standard	vSphere ROBO
N/A	Automated workload rebalancing and affinity rules	Manual workload balancing	Supports up to 25 virtual machines
N/A	Automated maintenance mode	Manual maintenance mode	Supports three editions of vSphere: Standard, Advanced, and Enterprise
N/A	One-click software updates	One-click software updates	Mainly used for either remote-office or Disaster Recovery sites
N/A	vGPU support for VxRail models supporting GPUs	It does not support vGPU	N/A
N/A	Supported dynamic nodes	Multistep drive replacement	N/A
N/A	Streamlined drive replacement	N/A	N/A
N/A	Supported active-active data center	N/A	N/A

Table 1.3 – The difference between the preceding supported vSphere licenses

Important Note
VMware vSphere Essentials and Essentials Plus are not supported with VxRail.

VxRail software 4.7 or above also supports a flexible vSAN license model. You can purchase VxRail Appliance with different editions of the vSAN license. This table shows an edition comparison:

Edition	Standard	Advanced	Enterprise	Enterprise Plus
Storage Policy Based Management	Supported	Supported	Supported	Supported
Virtual Distributed Switch	Supported	Supported	Supported	Supported
Software Checksum	Supported	Supported	Supported	Supported
All-Flash Hardware	Supported	Supported	Supported	Supported
iSCSI Target Service	Supported	Supported	Supported	Supported
QoS	Supported	Supported	Supported	Supported
Cloud-Native Storage	Supported	Supported	Supported	Supported
Deduplication and Compression	N/A	Supported	Supported	Supported
RAID-5/6 Erasure Coding	N/A	Supported	Supported	Supported
vRealize Operations with vCenter	N/A	Supported	Supported	Supported
Data-at-Rest Encryption	N/A	N/A	Supported	Supported
Stretched Cluster	N/A	N/A	Supported	Supported

Table 1.4 – VMware vSAN editions feature comparison

This table shows an edition comparison for vRealize Operations Advanced:

Edition	Standard	Advanced	Enterprise	Enterprise Plus
vSAN-aware WLB	N/A	N/A	N/A	Supported
Customizable Dashboards	N/A	N/A	N/A	Supported
Auto-Remediation	N/A	N/A	N/A	Supported
Troubleshooting Workflows	N/A	N/A	N/A	Supported
Automated Intent-Based Workload	N/A	N/A	N/A	Supported
Capacity Planning	N/A	N/A	N/A	Supported

Table 1.5 – VMware vRealize Operations Advanced edition feature comparison

A VMware vCenter Server Standard instance license is bundled with the VxRail cluster. VxRail Manager applies this license to the embedded VCSA during VxRail initialization.

> **Important Note**
>
> Transferring an embedded VCSA license to any vCenter Server is not supported. You need to prepare a new vCenter Server license for the external VCSA if you choose external VCSA during VxRail initialization.

The preceding tables help you choose the different types of VMware vSphere and vSAN licenses for VxRail Appliance based on your required functions and features.

Now that we understand VxRail Appliance and the different components and technologies used in it, let's move on and learn about its architecture next.

Dell VxRail architecture

The VxRail system is a turnkey solution that has been tested and validated by Dell EMC. Each VxRail node is built on a Dell PowerEdge server, and it includes the following hardware components:

- Intel Xeon Scalable processors (single, dual, or quad), up to 40 cores per processor, or an AMD EPYC processor with up to 64 cores.

- Up to 48 DDR4 **Dual In-Line Memory Modules** (**DIMMs**), it supports memory capacity ranging from 64 GB to 8,192 GB.

- A mirrored pair of BOSS SATA M.2 cards that are used to store the ESXi system on the node.

- A 10/25 GbE **Network Daughter Card** (**NDC**) used for VxRail's predefined network connections.

- If you purchase a VxRail Hybrid node, it includes a single **Solid State Drive** (**SSD**) for the cache tier and multiple **Hard Disk Drive** (**HDD**) disks for the capacity tier.

- If you purchase a VxRail All-Flash node, it includes a single SSD or NVMe for the cache tier and SAS SSD, SATA SSD, vSAS SSD, or NVMe for the capacity tier.

> **Important Note**
>
> Each VxRail Series can support different maximum software and hardware configuration, for example, the number of vSAN disk groups, the number of additional network adapters, or the total number of memory and CPU cores.

Now, we will look at a diagram for the VxRail 7.0.xxx platform:

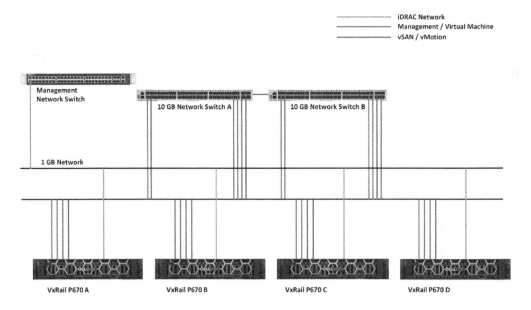

Figure 1.8 – A diagram of the VxRail 7.0.xxx platform

Figure 1.8 shows the following hardware components in this environment:

- There are four VxRail P670 Hybrid appliances (a minimum of four nodes is the recommended configuration). Each node installs an NDC with four 10 GB SFP+ ports. Two 10 GB ports are used for vSAN and vMotion networks; the other 10 GB ports are used for management and virtual machine networks.

- Two 10 GB network switches are used for VxRail's network connectivity.

- One 1 GB network switch is used for **Integrated Dell Remote Access Controller (iDRAC)** connection on each VxRail Appliance.

> **Important Note**
>
> The minimum initial configuration of a VxRail cluster is three nodes; these three nodes must be the same model.

Now, we will discuss the logical diagram for the VxRail 7.0.xxx platform. *Figure 1.9* shows the following software and hardware components in this environment:

- One VxRail 7.x cluster with four VxRail nodes, and there is a vSAN datastore across these four nodes

- One VxRail Manager virtual appliance

- One VCSA

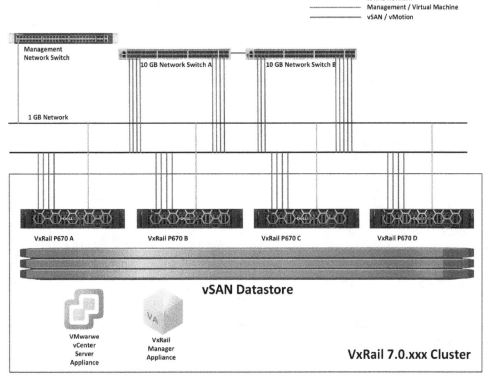

Figure 1.9 – The logical diagram for the VxRail 7.0.xxx platform

> **Important Note**
>
> In VxRail 7.0 or later, there is no VMware vCenter **Platform Service Controller** (PSC), and vRealize Log Insight deploys automatically into the VxRail cluster during initialization.

In *Figure 1.9*, there are four nodes with the same model (VxRail P670) connecting to the VxRail network during VxRail initialization; then it can automatically build the VxRail cluster, and vSAN features are enabled on the VxRail cluster. By default, there are two VxRail system virtual machines that will be deployed into the VxRail cluster; they are VxRail Manage and VCSA. You can easily build the VxRail cluster in **Dell EMC VxRail Deployment Wizard** when every requirement is ready. You can refer to *Figure 1.10* for the VxRail initialization:

Figure 1.10 – Welcome page of VxRail 7.0

> **Important Note**
>
> Please note that you can choose the VxRail software edition when you purchase VxRail Appliance: VxRail 4.7 or 7.0. VxRail 4.7 is shipped with VMware vSAN 6.7. VxRail 7.0 is shipped with VMware vSAN 7.0.

In this section, we understood the architecture of the VxRail system and the VxRail system's virtual machines. Next, let's look at its features.

Dell VxRail features

The VxRail system can deliver different features; it includes automatic deployment, flexible scale-up and scale-out, **Storage Policy-Based Management (SPBM)**, LCM, a single management dashboard, CDP, and a single-vendor, end-to-end ongoing support service. In day-one deployment, you can select the cluster type and then deploy the VxRail cluster and configure the SDS automatically. Compared to the traditional server storage architecture, it can minimize the deployment and configuration time. In *Figure 1.11*, you can see how you can select the type of the VxRail cluster in Dell EMC VxRail Deployment Wizard.

Figure 1.11 – Specifying the type of the VxRail cluster in Dell EMC VxRail Deployment Wizard

You can perform all operation tasks via VMware vCenter Server with the VxRail Manager plugin after the VxRail deployment is completed. In the VxRail Manager plugin, you can see all functions, such as **Updates**, **Certificate**, **Market**, and **Add VxRail Hosts**.

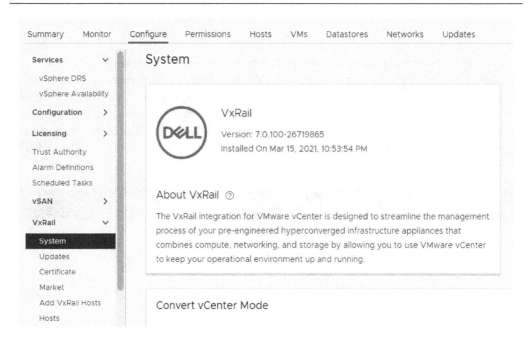

Figure 1.12 – VxRail System information

Storage Policy Based Management (**SPBM**) is a core feature in the VxRail cluster; the system administrator can define the availability, storage rules, and advanced policy rules. In the **Availability** tab, you can choose **No data redundancy**, **1 failure - RAID-1 (Mirroring)**, and **1 failure - RAID-5 (Erasure Coding)**. You can refer to *Chapter 4, Design of vSAN Storage Policies*, for more details.

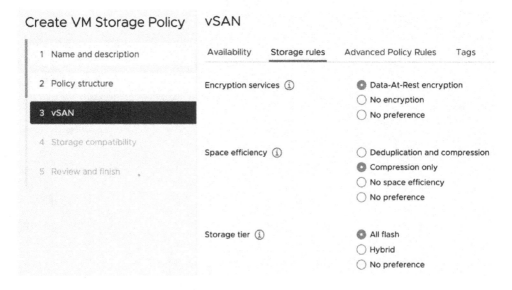

Figure 1.13 – Defining the storage rules of the virtual machine storage policy

VxRail's architecture supports flexible scale-up and scale-out. The compute and storage resources can be increased and rebalanced when adding a new node to the existing VxRail cluster (*Figure 1.14*). You can refer to *Chapter 5*, *Design of Cluster Expansion*, for more details:

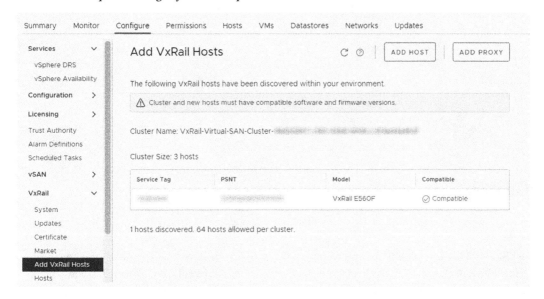

Figure 1.14 – The Add VxRail Hosts dashboard

When a system has been live for a few years, operating system and system patch upgrades are required. In a traditional server storage environment, the system administrator may spend a lot of time on the preparation and compatibility verification before the system upgrade.

LCM is one of the key features of VxRail Appliance; VxRail's one-click upgrade feature can easily handle LCM. The system administrator can easily upgrade the VxRail system with a single image that is prevalidated with Dell EMC and VMware. They can perform a one-click upgrade in vCenter Server with the VxRail Manager plugin (*Figure 1.15*). The single image includes VxRail Manager, VCSA, Dell PTAgent, VxRail Manager VIB, and VMware ESXi:

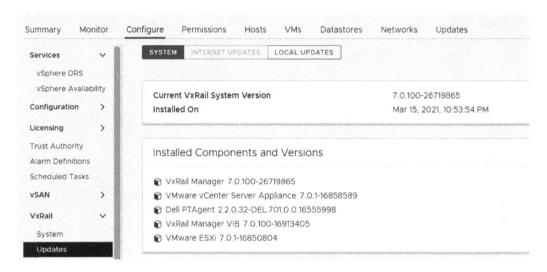

Figure 1.15 – The VxRail software update dashboard

Dell EMC RP4VM can provide data protection of the virtual machine with its point-in-time snapshots in a local VxRail cluster or across VxRail clusters between two different data centers. You can refer to *Chapter 9, Design of RecoverPoint for Virtual Machines on VxRail*, for more details.

VxRail's support service can deliver single-vendor support from Dell's support team. When the customer issues a request to Dell's support team, they can handle and resolve the Dell and VMware technical problems. This support service works with **Support Connect Gateway** (**SCG**); it includes three types of service support-level agreements, **Basic**, **ProSupport**, and **ProSupport Plus**. You can refer to *Figure 1.16* for details:

> **Note**
>
> SRS is replaced with SCG in VxRail 7.0.350 and above; it can deliver a single connectivity technology for Dell EMC products, such as server, data storage, and hyper-converged solutions. It can proactively monitor system health and create a case automatically over a secure connection.

Feature Comparison

	Basic	ProSupport	ProSupport Plus
Remote technical support	9x5	24x7	24x7
Covered products	Hardware	Hardware Software	Hardware Software
Onsite hardware support	Next business day	Next business day or 4hr mission critical	Next business day or 4hr mission critical
3rd party collaborative assistance		●	●
Self-service case initiation and management		●	●
Access to software updates		●	●
Proactive storage health monitoring, predictive analytics and anomaly detection with CloudIQ and the CloudIQ mobile app		●	●
Priority access to specialized support experts			●
Predictive detection of hardware failures			●
3rd party software support			●
An assigned Service Account Manager			●
Proactive, personalized assessments and recommendations			●
Proactive systems maintenance			●

Figure 1.16 – A comparison of Dell support services; this information is copyright of Dell Technologies (https://www.delltechnologies.com/asset/cs-cz/services/support/briefs-summaries/prosupport_enterprise_suite_brochure.pdf)

The customer can choose from the different types of support services to support their VxRail system based on their requirements.

Dell VxRail management

VxRail Manager is a core management virtual machine that is fully integrated with VMware vCenter Server via the VxRail Manager plugin for vCenter, as shown in *Figure 1.17*. The system administrator can perform all deployment and configuration activities from the vCenter HTML client:

Figure 1.17 – VxRail Manager plugin for vCenter

VxRail automates more than 200 configuration tasks and workflows, including initializing, configuring, building, and finishing. The following is a summary of each core configuration task:

- **Initializing:**

  - Deploy and configure vCenter Server.

  - Configure the **Domain Name System** (**DNS**) on vCenter Server.

  - Set up the management network on ESXi hosts.

  - Configure the time on ESXi hosts.

  - Configure the Syslog on ESXi hosts.

  - Configure the hostname on vCenter Server.

  - Create a user for VxRail management.

- **Configuring:**

  - Register ESXi hosts with vCenter Server.

  - Create the storage policies on vCenter Server.

  - Set up the hostnames on ESXi hosts.

  - Rename the vCenter Server network on ESXi hosts.

  - Set up **Network Interface Card** (**NIC**) bonding on ESXi hosts.

  - Set up vSAN, vMotion, and virtual machine networks on ESXi hosts.

- Set up NIC teaming on ESXi hosts.

- Set up DNS on ESXi hosts.

- Restart the loudmouth service on ESXi hosts.

- Set up the clustering for ESXi hosts.

- **Building:**

  - Restart the loudmouth service.

  - Accept the vCenter Server **End User License Agreement (EULA)**.

  - Create the vCenter Server database.

  - Create the configuration in vCenter Server.

  - Configure the root account on ESXi hosts.

  - Initialize vCenter **Single Sign-On (SSO)**.

  - Start vCenter Server.

- **Finishing:**

  - Rename the database.

  - Configure the root account on ESXi hosts.

  - Copy files to the vSAN datastore.

Compared to traditional infrastructure architecture, the preceding configuration tasks are executed manually. Now VxRail installation is an automated deployment.

VxRail LCM can deliver internet and local software updates. The LCM upgrade will provide an estimated time to complete the upgrade, warnings and recommended actions, and some components that will be updated (*Figure 1.18*):

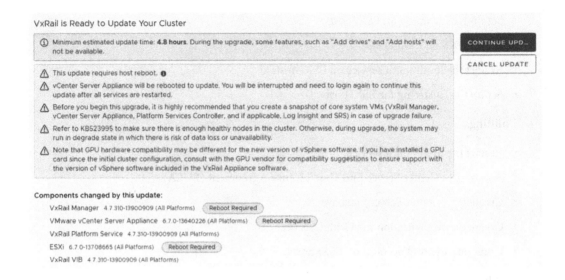

Figure 1.18 – VxRail LCM update dashboard

When the VxRail administrator executes the maintenance tasks on the VxRail system, for example, scale-out, hardware placement, or software package upgrades, they can perform these tasks via the VxRail Manager plugin for vCenter.

Dell VxRail documentation and resources

In this last section, we will discuss where you can access the VxRail documentation and resources, which include the Dell Technologies support website, Dell Technologies SolVe Online, and VMware documentation.

As you can see in *Figure 1.19*, you can access the VxRail software, knowledge base, and documentation on the **Dell Technologies support website** if you are a Dell partner or employee. The access link is at https://www.dell.com/support/home/en-us/product-support/product/vxrail-appliance-series/overview:

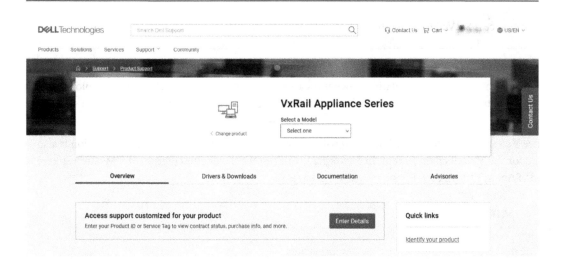

Figure 1.19 – Dell Technologies support site

> **Important Note**
>
> Please note that some VxRail software and documentation can only be accessed by Dell Technologies employees.

Dell Technologies SolVe Online is a web-based tool (*Figure 1.20*) used by Dell Technologies employees, partners, and customers. This tool is used to generate detailed installation and configuration procedures for all Dell hardware and software products, for example, VxRail Appliance and RP4VM.

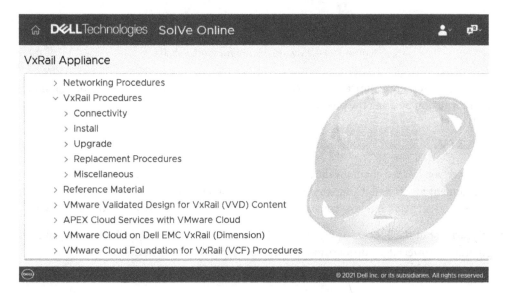

Figure 1.20 – Dell Technologies SolVe Online

You can access SolVe Online with this link: `https://solve.dell.com/solve/home`.

> **Important Note**
> Please note that some installation procedures are only accessed by Dell Technologies employees and certified partners. Dell Technologies SolVe also has an interactive edition, SolVe Desktop.

Since VxRail Appliance is fully integrated with VMware products, the VMware documentation is also valid for the VxRail environment. In *Figure 1.21*, you can see the documentation of VMware vSphere and vSAN at `https://docs.vmware.com/`:

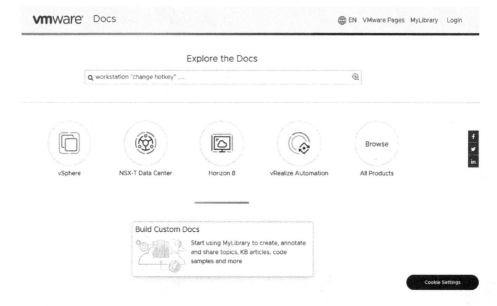

Figure 1.21 – VMware documentation website

If you are looking for any technical or knowledge support about VxRail Appliance, you can find the relevant information by using the preceding documentation and resources.

Summary

In this chapter, you learned about the different series of the VxRail Appliance 7.x system, architecture, and management for VxRail. Compared to the traditional architecture of server and storage, you learned how VxRail helps and how you can offload the operation and configuration on VxRail.

You will learn about the benefits of VxRail Appliance in the next chapter, including VxRail on the 15th-generation PowerEdge portfolio and the different types of VxRail nodes.

Questions

The following are a short list of review questions to help reinforce your learning and help you identify areas which require some improvement.

1. Which license is not bundled with VxRail Appliance?

    A. VMware vCenter Server Appliance

    B. VMware vSAN

    C. Dell EMC RecoverPoint for Virtual Machines

    D. VMware vRealize Log Insight

    E. VMware Cloud Foundation

2. Which VxRail Series does not exist?

    A. E Series

    B. P Series

    C. S Series

    D. K Series

    E. V Series

3. If the customer is planning to deploy some virtual machines with high performance, as a system consultant, which VxRail series should you propose to the customer?

    A. VxRail P670F

    B. VxRail V670F

    C. VxRail E660

    D. VxRail S670

    E. VxRail G560F

4. Which of these is an unsupported configuration of VxRail?

    A. VxRail V Series All-Flash

    B. VxRail E Series Hybrid

    C. VxRail V Series NVMe

    D. VxRail S Series Hybrid

    E. VxRail G Series Hybrid

5. What are the software components that exist in the VxRail system?

    A. VMware Site Recovery Manager

    B. VMware Cloud Foundation

    C. VMware vRealize Operations

    D. VMware vSphere Replication

    E. VMware Tanzu

6. Which vSAN license editions can be supported with RAID-5/6 Erasure Coding?

    A. vSAN Standard

    B. vSAN Advanced

    C. vSAN Enterprise

    D. vSAN Enterprise Plus

    E. vSAN ROBO

7. Which vSphere license edition is used in a remote office environment?

    A. vSphere Standard

    B. vSphere Advanced

    C. vSphere Enterprise

    D. vSphere Enterprise Plus

    E. vSphere ROBO

8. Which of these is not a feature of VxRail?

    A. Continuous data protection

    B. One-click upgrade

    C. Storage policy-based management

    D. Data-at-rest encryption

    E. Storage replication

9. Which of these can be supported on NVMe configuration?

    A. VxRail V Series

    B. VxRail E Series

    C. VxRail P Series

    D. VxRail S Series

E.   VxRail G Series

F.   VxRail D Series

10. Which VxRail service-level agreement includes Service Account Manager?

A.   Basic

B.   Enterprise

C.   Enterprise Plus

D.   ProSupport

E.   ProSupport Plus

F.   None of the above

11. Which resources do you use to generate installation procedures for the VxRail system?

A.   Dell Technologies support website

B.   Dell Technologies SolVe Online

C.   My VMware support website

D.   VMware documentation website

E.   Dell Technologies SolVe Desktop

12. Which VxRail system virtual machines do not exist on the VxRail 7.0. system?

A.   vRealize Log Insight

B.   VxRail Manager

C.   vCenter Server Appliance

D.   vCenter Server Platform Controller

E.   vRealize Operations Manager

2

Benefits of Hyper-Converged Infrastructure

The previous chapter provided an overview of the VxRail Appliance 7.x system. You learned about the architecture, features, management, and documentation resources for VxRail, such as automated deployment, one-click upgrade, and storage policy-based management. You learned how to help and offload the operation and configuration of VxRail in comparison with the traditional server and storage architecture. The VxRail 7.x platform runs on the latest generation of Dell PowerEdge hardware, the 15th generation. Compared to the previous generation of Dell PowerEdge hardware, the performance and configuration are improved in the latest version.

VxRail Appliance is a **Hyper-Converged Infrastructure** (**HCI**) that is a self-contained infrastructure platform, and you can select different hardware components (such as CPU processors, memory, network connectivity, a network daughter card, GPUs, NVMe, and all-flash disk devices) to build your environment based on your hardware and software requirements. If you don't have hardware and software requirements in the initial phase, you can build your VxRail cluster with the minimum configuration (that is, three VxRail nodes with the same model and configuration). You can easily scale up or scale out your VxRail cluster when you need to upgrade or expand the resources on the VxRail platform in the future. The scaling feature is one key benefit of VxRail Appliance. VxRail 15th generation hardware enhances some features and cluster deployment, and you will learn about the benefits of this latest generation of HCI in this chapter.

This chapter includes the following main topics:

- What's new in VxRail 15th generation?
- VxRail nodes with vSAN
- VxRail dynamic nodes
- VxRail satellite nodes

What's new in VxRail 15ᵗʰ generation?

VxRail supports various use cases, such as big data analytics, **virtual desktop infrastructure** (VDI), remote office, and Brand Office. In VxRail 7.x, support to use your own VMware vSphere and vSAN license or obtain VMware licenses from Dell Technologies is available. VxRail provides the following benefits:

- It can deliver a hybrid cloud with VMware Cloud Foundation on VxRail, including Dell Technologies Cloud Platform, VMware Cloud on Dell EMC, and **VMware Validated Designs** (**VVD**).

- It can deliver cloud-based management and analytics services for the VxRail platform.

- It supports the services of active-active data centers and active-passive solutions.

- It supports end-to-end lifecycle management and support.

The following table shows a summary of improvements on four new platforms:

	CPU	Memory	Connectivity	Storage
E660	New CPU processors up to 40 cores.	Next-generation Intel Optane persistent memory.	Supports PCIe Generation 4.	Supports PCIe Generation 4 NVMe cache devices.
E660F	N/A	Supports up to 8 TB of memory.	SAS HBA with x16 SAS lanes.	Supports NVMe cache drives on VxRail V and S Series.
E670F	N/A	Supports up to 8 TB of 2ⁿᵈ generation Intel Optane persistent memory.	Supports quad-port 25 Gb OCP 3.0 networking.	Supports 12 TB NL-SAS capacity drives on VxRail S Series.
V670F	N/A	N/A	Supports hot-pluggable **Boot Optimized Storage Solution** (**BOSS**).	Supports the additional four capacity disk slots on VxRail P Series for up to 184 TB of storage.

Table 2.1 – A summary of improvements on each node in VxRail 15th generation

VxRail 14th generation (prior to VxRail 7.0.240) only supported one type of VxRail cluster deployment (Standard Cluster) configured with physical compute and storage resources on each node. The compute and storage resources can be increased at the same time when you add one or more nodes into the existing VxRail cluster. VxRail v4.7.100 supports the vSAN two-node cluster with a direct-connect configuration, but this configuration does not support scaling. We will discuss the details of the vSAN two-node cluster in *Chapter 6, Design of vSAN 2-Node Cluster on VxRail*. VxRail also supports vSAN Stretched Clusters; we will discuss the details in *Chapter 7, Design of Stretched Cluster on VxRail*. VxRail 15th generation (VxRail 7.0.240 or later) supports another type of VxRail cluster deployment, **Dynamic Node Cluster**. From VxRail 7.0.300, it supports **VxRail satellite nodes**. When initializing a VxRail 7.x system, the Dynamic Node Cluster can be selected in the VxRail Deployment Wizard.

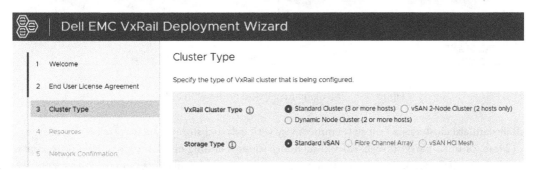

Figure 2.1 – Dell EMC VxRail Deployment Wizard

The following sections will provide an overview of and discuss the fundamental differences between each VxRail node, that is, VxRail nodes with vSAN, VxRail dynamic nodes, and VxRail satellite nodes.

VxRail nodes with vSAN

The VxRail Standard Cluster is one type of VxRail cluster deployment; it requires a minimum of three nodes with the same model in the VxRail cluster, which all VxRail hardware models support. Each VxRail node includes cache disks (Flash or NVMe) and capacity disks for this cluster type. In *Figure 2.2*, there are four nodes in the VxRail cluster, and each node consists of a **Disk Group** with one cache disk and two capacity disks. It can automatically build the VxRail cluster and create a local vSAN datastore as primary storage after the deployment is complete.

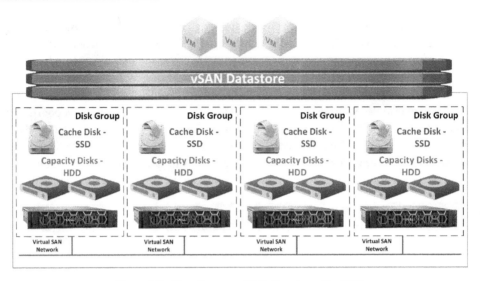

Figure 2.2 – The diagram of VxRail nodes with vSAN

VxRail standard clusters also support connectivity with external storage, such as Dell EMC storage, HCI Mesh, or third-party storage. All VxRail nodes support adding one Fibre Channel HBA except the VxRail D and G Series. You can set up the connectivity of the VxRail cluster and secondary storage with Fibre Channel or iSCSI Channel.

VxRail standard cluster with external storage

In *Figure 2.3*, there is a VxRail standard cluster with four nodes connected with a Dell storage array over Fibre Channel. In this configuration, you can move the virtual machines across the vSAN datastore and SAN datastore with vSphere Storage vMotion. The virtual machine can be shared between the primary and secondary storage resources.

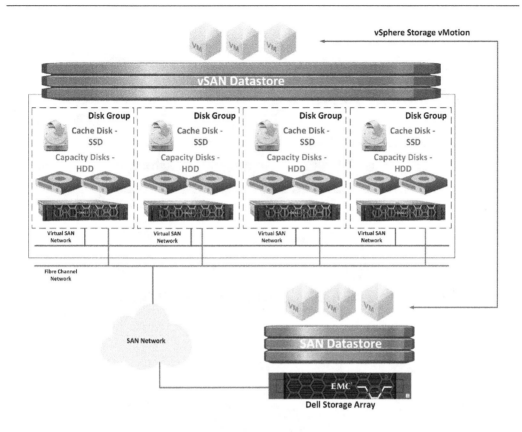

Figure 2.3 – VxRail standard cluster with external storage

Figure 2.3 gives an overview of a VxRail standard cluster with external storage.

> **Important Note**
>
> VxRail Lifecycle Management (one-click upgrade) does not include the firmware upgrade of an external storage array.

We will discuss the VxRail standard cluster with vSAN HCI Mesh in the next section.

VxRail standard cluster with vSAN HCI Mesh

From VxRail 7.0.100, VxRail supports vSAN HCI Mesh. With this feature, we have a local vSAN datastore on a VxRail cluster that can be shared with other VxRail clusters. In *Figure 2.4*, there are two VxRail standard clusters (VxRail Cluster A and Cluster B) with four nodes. The local vSAN datastore on VxRail Cluster B can be shared with VxRail Cluster A.

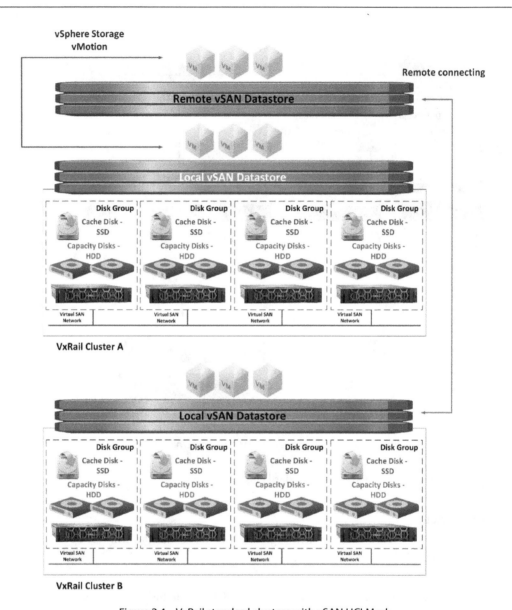

Figure 2.4 – VxRail standard clusters with vSAN HCI Mesh

This configuration lets you move the virtual machines across the local vSAN datastore and remote vSAN datastore with vSphere Storage vMotion. The virtual machine can be shared between the local and remote storage resources.

If you want to enable a VxRail standard cluster with vSAN HCI Mesh, please prepare the following:

- VxRail software 7.0.100 or later.

- A single vCenter instance to manage each VxRail cluster that enables vSAN HCI Mesh.

- A VMware vSAN Enterprise or Enterprise Plus license is required for the VxRail cluster to share the vSAN datastore.

- A maximum of 1 millisecond round-trip time for VxRail clusters between all VxRail nodes in the cluster.

VxRail standard cluster deployment is suitable for most common scenarios; this cluster type has the following features:

- It can deliver a single management dashboard of computing and storage resources. And it is easy to handle the life cycle management in the HCI platform.

- It can deliver a simple deployment and user-friendly operation.

- The storage resources are not dependent on secondary storage.

- For system scale-out, the compute and storage resources can be increased simultaneously. It can deliver scalability and **high-availability** (**HA**) requirements.

- Support configuring the external storage resources on the VxRail cluster as a secondary storage array.

In some scenarios, VxRail standard cluster deployment may not be suitable, as follows:

- The storage requirements are very large.

- The environment is a remote office or Brand Office.

- The service-level agreement and HA of applications are very high. For example, deploying a VxRail Stretched Cluster to protect all application virtual machines.

- The customer wants to increase the storage resources for system scale-out, and the compute resources are not required.

Thanks to the examples in *Figure 2.3* and *Figure 2.4*, you understand the benefits of VxRail nodes with vSAN and which scenarios are suitable for this type of cluster deployment.

In VxRail 14th generation, you can only use the VxRail standard cluster deployment. We will discuss another type of VxRail cluster in the next section, VxRail dynamic nodes.

VxRail dynamic nodes

The VxRail dynamic node cluster is a type of VxRail cluster deployment that requires a minimum of two nodes with the same model in the VxRail cluster, supported by three VxRail hardware models, E660F, P670F, and V670F. These nodes are built on Dell PowerEdge 15th-generation servers. VxRail dynamic nodes do not include any storage resources and only consist of the compute resources. However, these nodes can also deliver all the benefits of the VxRail Appliance system, except the storage resources. If you choose this type of cluster deployment, you need to connect to an external storage array. The external storage supports various options, including Dell PowerStore, Dell PowerMax, and Dell Unity XT. In *Figure 2.5*, there is a VxRail dynamic node cluster with four nodes connected to an external storage array. In this type of cluster deployment, the connectivity of the VxRail dynamic node and external storage is in **Fibre Channel** (**FC** in the diagram).

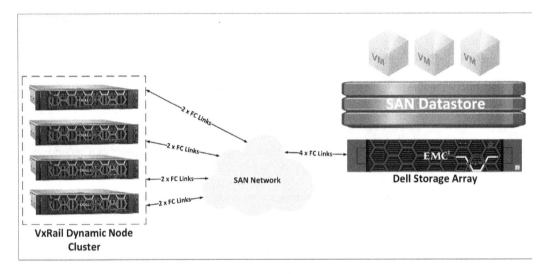

Figure 2.5 – VxRail dynamic node cluster

If you choose to use a VxRail dynamic node cluster deployment, you need to consider the following:

- If you want to scale out the system and you need to increase the compute resources but storage resources are not required, a VxRail dynamic node is a good option.

- If you plan to reuse the existing storage resources as the primary storage.

- There is Dell-certified storage as the primary storage for the VxRail dynamic node cluster in your environment.

- When you choose the supported storage for the VxRail dynamic node, please check the VxRail Support Matrix for compatibility.

- For HA and redundancy on each VxRail dynamic node, I suggest installing two Fibre Channel dual-port adapters per VxRail dynamic node. The Fibre Channel adapter supports Emulex and QLogic models.

- Lifecycle management does not include the primary storage resources in a single management dashboard of computing resources.

- Make sure there are two Fibre Channel switches for the connections of each VxRail dynamic node.

> **Important Note**
> The VMware vSAN license is not required on each VxRail dynamic node if you plan to deploy the VxRail dynamic node cluster.

We will discuss the VxRail dynamic node cluster with vSAN HCI Mesh in the next section.

VxRail dynamic node cluster with vSAN HCI Mesh

VxRail dynamic node clusters also support vSAN HCI Mesh. In this feature, the local vSAN datastore on a VxRail cluster can be shared with VxRail dynamic node clusters. Now we will discuss two vSAN HCI Mesh scenarios.

Scenario 1

In *Figure 2.6*, VxRail Cluster A has four nodes and a local vSAN datastore across this VxRail cluster. This local vSAN datastore is shared with the VxRail dynamic node cluster. In this configuration, the virtual machines on the VxRail dynamic node cluster can use the storage resources on VxRail Cluster A. You can move the virtual machines on the VxRail dynamic node cluster into either VxRail Cluster A or the remote vSAN datastore using vSphere Storage vMotion.

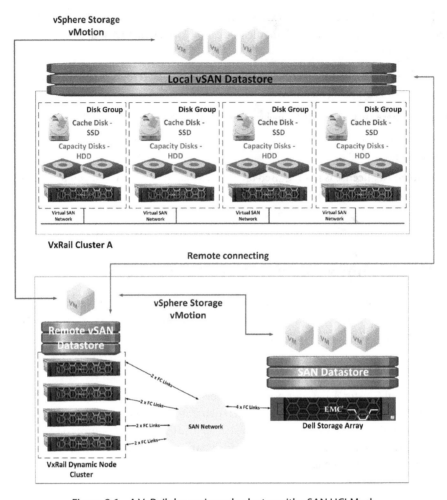

Figure 2.6 – A VxRail dynamic node cluster with vSAN HCI Mesh

If you choose to use a VxRail dynamic node cluster with vSAN HCI Mesh, you need to consider the following prerequisites:

- VxRail 7.0.100 or later.

- A VMware vSAN Enterprise or Enterprise Plus license is required for the VxRail cluster shared with the vSAN datastore.

- When you choose the supported storage for the VxRail dynamic node, please check the VxRail Support Matrix for compatibility.

- If the vSAN datastore is already shared with other clusters, ensure it only supports the maximum of five clusters connected to this vSAN datastore.

- Both the VxRail dynamic node cluster and the VxRail cluster sharing its vSAN datastore need to be managed by the same vCenter instance.

- The round-trip time latency between the dynamic nodes and the VxRail cluster sharing its vSAN datastore must be less than 5 milliseconds.

Scenario 2

In *Figure 2.7*, each VxRail cluster has four nodes and a local vSAN datastore across this VxRail cluster. This local vSAN datastore is shared with VxRail Dynamic Node Cluster A and Dynamic Node Cluster B.

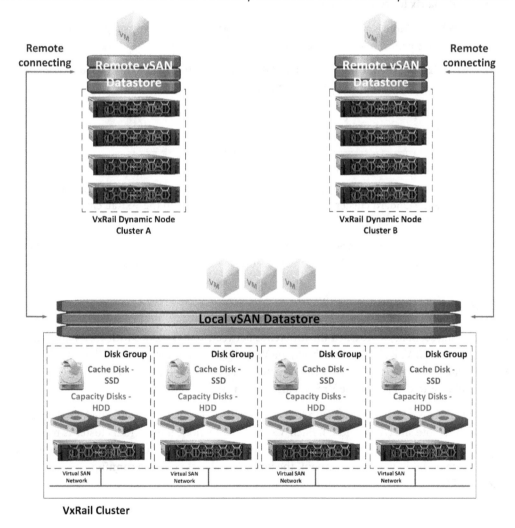

Figure 2.7 – The two VxRail dynamic clusters with vSAN HCI Mesh

If you choose two VxRail dynamic node clusters with vSAN HCI Mesh, you need to consider the following requirements:

- VxRail 7.0.240 or later.

- A VMware vSAN Enterprise or Enterprise Plus license is only required for the VxRail cluster shared with the vSAN datastore. The vSAN license is not required on each VxRail dynamic node cluster.

- When you choose the supported storage for the VxRail dynamic node, please check the VxRail Support Matrix for compatibility.

- If the vSAN datastore is already shared with other clusters, ensure it only supports the maximum of five clusters connected to this vSAN datastore.

- Both the VxRail dynamic node cluster and the VxRail cluster sharing its vSAN datastore need to be managed by the same vCenter instance.

- The round-trip time latency between the dynamic nodes and the VxRail cluster sharing its vSAN datastore must be less than 5 milliseconds.

> **Important Note**
>
> In VxRail 7.0.240 and above, the vSAN license is not required on each VxRail dynamic node cluster or VxRail standard cluster if vSAN HCI Mesh is enabled on the VxRail cluster.

Thanks to the preceding two scenarios in *Figure 2.6* and *Figure 2.7*, you now understand the benefits of VxRail dynamic nodes and which scenarios are suitable for this type of cluster deployment.

VxRail satellite nodes

In the preceding sections, we learned the difference between VxRail nodes with vSAN and VxRail dynamic nodes. From VxRail 7.0.320 or above, the VxRail satellite node is available. This node is a type of VxRail node and it supports lifecycle management through VxRail Manager. But it only requires a single IP address to connect to the VxRail cluster in an HQ data center.

In *Figure 2.8*, there is a VxRail cluster with four nodes in the HQ data center and one VxRail satellite node in each remote site (A, B, and C). Each VxRail satellite node and VxRail cluster is managed by a single vCenter instance through a **Wide Area Network** (**WAN**) in the HQ data center. If you deployed these nodes in your environment, the virtual machines could be running on local RAID storage (PERPC H755 controller) or secondary storage.

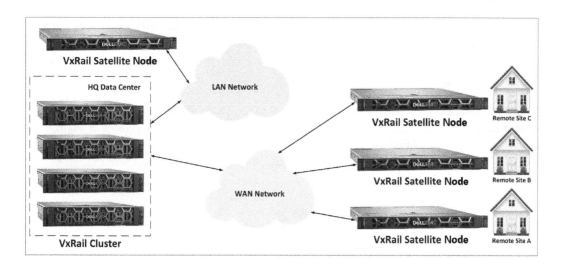

Figure 2.8 – VxRail satellite nodes

If you plan to deploy VxRail satellite nodes in your environment, you need to consider the following prerequisites:

- VxRail 7.0.320 or later is required to support VxRail satellite nodes.

- VxRail satellite nodes do not support vSAN HCI Mesh.

- VxRail satellite nodes are supported in a single instance, and it cannot support node expansion.

- VxRail satellite nodes are used in remote sites or edge locations.

- If a VxRail cluster with a vSAN datastore is already deployed in your environment, then you can add the satellite nodes into the VxRail cluster for the management of each satellite node.

- Data and applications are not protected from node failure.

- The PERC H755 controller provides local RAID protection, including RAID 0, 1, 5, 6, 10, 50, and 60.

- VxRail satellite nodes support the VMware Standard edition or Enterprise Plus edition.

- VxRail satellite nodes only support customer-managed vCenter Server; embedded vCenter Server is not supported.

> **Important Note**
> The VMware vSAN license is not required on each VxRail satellite node if you plan to deploy VxRail satellite nodes.

Thanks to *Figure 2.8*, you have had an overview of VxRail satellite nodes. Now we will discuss an example of VxRail satellite nodes in the next section.

Scenario

In *Figure 2.9*, you can see there is a VxRail management cluster in the primary data center. The VxRail management cluster is used as a VxRail with vSAN cluster, which is managed by a customer-managed vCenter. The single VxRail management cluster can manage many VxRail satellite nodes in different locations. The lifecycle management of VxRail satellite nodes is supported in this scenario.

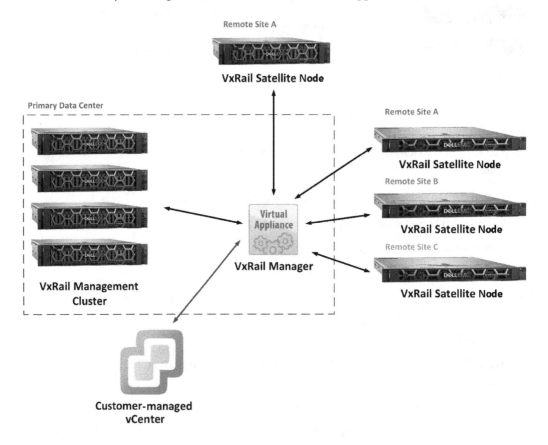

Figure 2.9 – An example configuration of VxRail satellite nodes

You now understand the benefits of VxRail satellite nodes and which scenarios are suitable for this type of cluster deployment. This table shows a summary of each kind of VxRail deployment:

	VxRail Node with vSAN	VxRail Dynamic Node	VxRail Satellite Node
Hardware model	It supports all VxRail models	It supports VxRail E660F, P670F, and V670F	It supports VxRail E660, E660F, and V670F
Boot device	BOSS with RAID-1	BOSS with RAID-1	BOSS with RAID-1
Primary storage	VMware vSAN	VMware vSAN HCI Mesh or Dell EMC storage	Local RAID storage PERC H755
Secondary storage	VMware vSAN HCI Mesh, Dell EMC storage, or third-party SAN	VMware vSAN HCI Mesh, Dell EMC storage, or third-party SAN	Dell EMC storage or third-party SAN
Scaling	Scale from 2 to 64 nodes	Scale from 2 to 96 nodes	It is not supported
VxRail Lifecycle Management	It is supported	It is supported but not included in the storage	It is supported
VMware vSAN license	It supports all VMware vSAN editions	No VMware vSAN license required	No VMware vSAN license required
Used scenario	All common cases	Independent scaling of computing resources	Edge

Table 2.2 – A summary of each kind of VxRail deployment

In this section, you got an overview of VxRail satellite nodes and their advantages.

> **Important Note**
> In a VxRail node with vSAN, the boot device is a single drive only in G Series. In the VxRail dynamic node, the vSAN license is not required in VxRail software 7.0.240 or later.

Summary

In this chapter, you learned about the new features and hardware in VxRail 15th generation, including VxRail cluster types, vSAN dynamic nodes, and VxRail satellite nodes. You also explored the benefits of each kind of VxRail deployment and the scenarios in which each deployment is suitable.

In the next chapter, you will learn how to design a vCenter server for the VxRail 7 system, including an internal vCenter server with external DNS and internal DNS, an external vCenter server with external DNS, and an internal vCenter server with a customer-supplied virtual distributed switch.

Questions

The following is a short list of review questions to help reinforce your learning and help you identify areas that require some improvement:

1. Which VxRail cluster types can be configured in the VxRail Deployment Wizard?

    A. VxRail standard cluster

    B. VxRail stretched cluster

    C. VxRail dynamic node cluster

    D. VxRail satellite node cluster

    E. VxRail two-node cluster

2. Which storage types can be configured in the VxRail Deployment Wizard?

    A. Standard vSAN

    B. iSCSI Channel array

    C. vSAN HCI Mesh

    D. Dell EMC storage array

    E. Fibre Channel array

    F. All of the above

3. Which VxRail software version can vSAN HCI Mesh support?

    A. VxRail 4.5.xxx

    B. VxRail 4.7.xxx

    C. VxRail 7.0.xxx

    D. VxRail 4.7.100

    E. VxRail 7.0.100

    F. None of the above

4. Which VMware vSAN licenses are required for the VxRail cluster to share the vSAN datastore when vSAN HCI Mesh is enabled?

    A. VMware vSAN Standard edition

B. VMware vSAN Advanced edition

C. VMware vSAN Enterprise edition

D. VMware vSAN Enterprise Plus edition

E. All of the above

5. Which VxRail feature is not supported on the VxRail dynamic node cluster?

A. VMware vSphere vMotion

B. VMware vSphere HA

C. VMware vSphere DRS

D. Define a vSAN datastore

E. Connect to a remote vSAN datastore

6. What scenarios are not recommended to deploy a VxRail standard cluster?

A. The storage resource requirements are very large.

B. When you scale out the system, the compute and storage resources can be increased simultaneously.

C. The environment is a remote office or brand office.

D. The environment is a primary data center.

E. The external storage resources on the VxRail cluster need to be configured as a secondary storage array.

7. Which VxRail cluster types can be supported with vSAN HCI Mesh?

A. VxRail two-node cluster

B. VxRail three-node cluster

C. VxRail dynamic node cluster

D. VxRail satellite node cluster

E. VxRail stretched cluster

F. All of the above

8. Which VxRail software version can support VxRail satellite nodes?

A. VxRail 4.5.xxx

B. VxRail 4.7.xxx

C. VxRail 7.0.xxx

D. VxRail 4.7.100

E.   VxRail 7.0.300

F.   All of the above

9.   Which configuration is only supported with an external vCenter Server?

A.   VxRail standard cluster

B.   VxRail dynamic node cluster

C.   VxRail satellite node

D.   VxRail two-node cluster

E.   VxRail stretched cluster

F.   None of the above

10. Which VxRail node can support local RAID storage PERC H755?

A.   VxRail node with vSAN

B.   VxRail satellite node

C.   VxRail dynamic node

D.   VxRail D560

E.   VxRail G560

F.   All of the above

11. Which use cases are recommended for VxRail satellite nodes?

A.   Active-active data centers

B.   Disaster recovery solutions

C.   Remote office

D.   Edge

E.   Independent scaling of computing resources

F.   None of the above

12. A customer plans to deploy the new VxRail Appliance version in their environment; the requirements are listed here:

- Reuse the existing storage array.

- Independent scaling of computing resources.

- The life cycle management of storage is not required.

- The Fibre Channel connection is supported.

Which diagram is recommended configuration for this use case?

A.   VxRail standard cluster

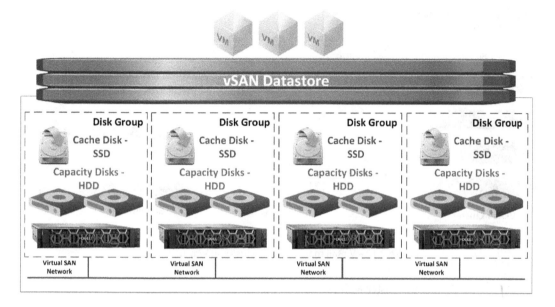

Figure 2.10 – VxRail standard cluster environment

B.   VxRail dynamic node cluster

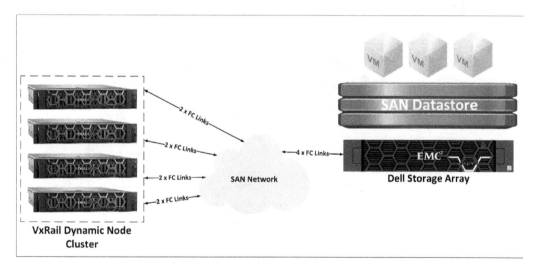

Figure 2.11 – VxRail dynamic node cluster environment

C. VxRail management cluster

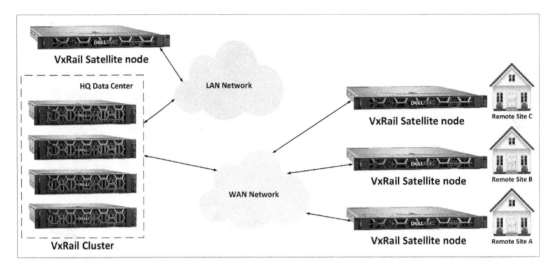

Figure 2.12 – VxRail management cluster environment

D. VxRail standard cluster with external storage array

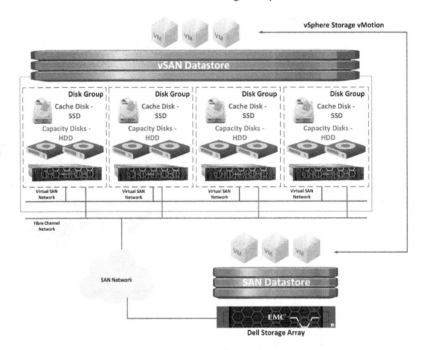

Figure 2.13 – VxRail standard cluster with external storage array environment

E. VxRail dynamic node cluster with vSAN HCI Mesh

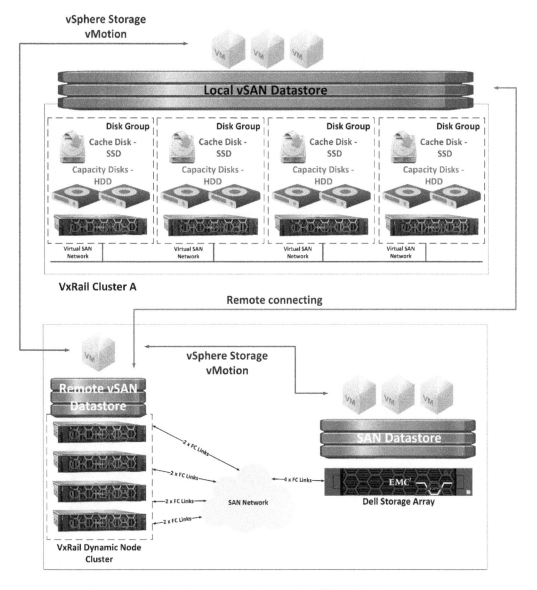

Figure 2.14 – VxRail dynamic node cluster with vSAN HCI Mesh environment

Part 2: Design of the VxRail Appliance 7.x System

In this section, the reader will learn about the design of VxRail deployment options, the best practices for vSAN storage policies, cluster expansion, and vSAN 2-Node clusters in VxRail.

This part of the book comprises the following chapters:

3

Design of vCenter Server

In the previous chapter, you learned about the new features and hardware in the 15th-generation VxRail, including VxRail cluster types, vSAN dynamic nodes, and VxRail satellite nodes. When you plan to deploy the VxRail system in your environment, you can choose different kinds of VxRail deployments. You learned the advantages and disadvantages of each kind of VxRail node and what use case is suitable for each VxRail node.

VMware vCenter Server is an essential component of the VxRail system. VxRail Manager is integrated into VMware vCenter Server with the VxRail Manager plugin for vCenter. In the 14th-generation VxRail, you can choose the internal VMware vCenter Server (embedded vCenter Server) and external VMware vCenter Server (customer-supplied vCenter Server) for VxRail deployment in the Dell EMC VxRail deployment wizard. VxRail deployment depends on the design of the VxRail cluster configured with VMware vCenter Server, the **Domain Name Server (DNS)**, and the VMware **Virtual Distributed Switch (VDS)** in the customer environment. A VxRail cluster can be added to the internal VMware vCenter Server or external VMware vCenter Server during the initial configuration. Four VxRail deployment configurations are available when you install a VxRail system in the customer environment, that is, internal vCenter with an external DNS, internal vCenter with an internal DNS, external vCenter with an external DNS, and external vCenter with a customer-supplied VDS. This chapter explores VxRail deployment with the different types of vCenter Server deployments.

This chapter includes the following main topics:

- Internal vCenter Server with an external DNS
- Internal vCenter Server with an internal DNS
- External vCenter Server with an external DNS
- External vCenter Server with a customer-supplied VDS

Internal vCenter Server with an external DNS

This section will discuss VxRail deployment with the internal vCenter Server and an external DNS, and go through a scenario for this VxRail deployment.

Overview

In *Figure 3.1*, you can see two clusters connected to an external DNS server, **VxRail Cluster** and **vSphere Cluster**, in the VMware environment. The vSphere cluster is a standard cluster that is managed with vCenter Server, and this VxRail configuration is the VxRail standard deployment option. VxRail Manager and the internal vCenter Server (embedded vCenter Server) are installed on **VxRail Cluster**. Each VxRail node and all virtual machines (VxRail Manager, vCenter Server, and the other virtual machines) are configured in the customer DNS server host records. All the virtual network port groups are configured on VDS in the VxRail cluster.

If you choose the VxRail deployment with the internal vCenter Server and an external DNS, you need to consider the following:

- The internal vCenter Server only supports managing VxRail clusters; non-VxRail clusters are not supported.

- The internal vCenter Server supports the VxRail standard cluster and vSAN Stretched Cluster; vSAN two-node cluster deployment is not supported.

- The VMware vCenter Server Standard edition license is bundled on the internal vCenter Server.

- In VxRail 4.7.xxx, the internal VMware vCenter Server is deployed with an external VMware **Platform Services Controller** (**PSC**).

- In VxRail 7.0.xxx, the internal VMware vCenter Server is deployed with an embedded VMware PSC.

- VxRail lifecycle management includes the software upgrade of the internal vCenter Server.

- The internal vCenter Server also supports the management of multiple VxRail clusters.

Figure 3.1 shows an example environment of the internal VMware vCenter Server with an external DNS:

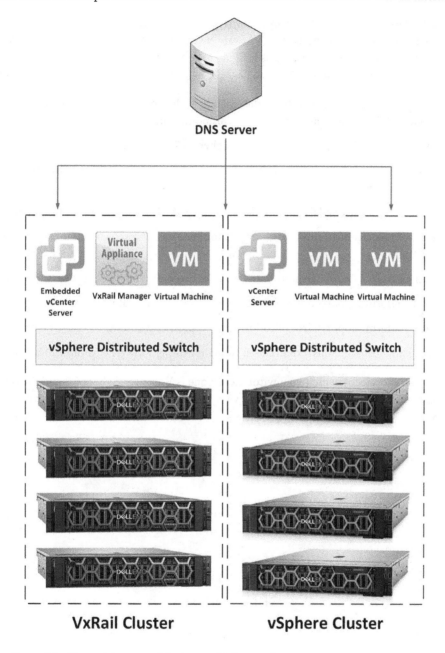

Figure 3.1 – The architecture of the internal VMware vCenter Server with an external DNS

Now, we will discuss a scenario for this type of deployment.

Scenario

In this scenario, the customer plans to deploy a new **Hyper-Converged Infrastructure** (**HCI**) platform in their VMware environment. In the existing VMware environment, the running platform is VMware vCenter 6.7 and vSphere 6.7. They expect this HCI platform to extend to a disaster recovery solution at the remote site for future planning. They also have the following requirements for this deployment:

- The VxRail cluster is separated from the existing VMware environment. The management panel of the VxRail cluster and non-VxRail cluster must be separated.

- The target version of VMware vCenter and vSphere is 7.0 in the VxRail cluster.

- The customer does not provide the optional VMware vCenter Server license for this VxRail deployment.

- They are using the existing DNS and **Active Directory** (**AD**) server for this VxRail deployment.

- The lifecycle management must be supported for hardware and software components on this HCI platform.

- The HCI platform supports extending into the disaster recovery solution or active-active solution.

- The HCI platform supports the vSphere Network I/O Control feature.

- The HCI platform supports deduplication and compression features.

- Migrating workloads into this VxRail cluster from the existing VMware environment is supported.

In *Figure 3.2*, you can see there are two environments; one is VMware vSphere 6.7 and the other is the VxRail cluster. The VxRail cluster is deployed with four VxRail P570F nodes. Both environments share the existing DNS:

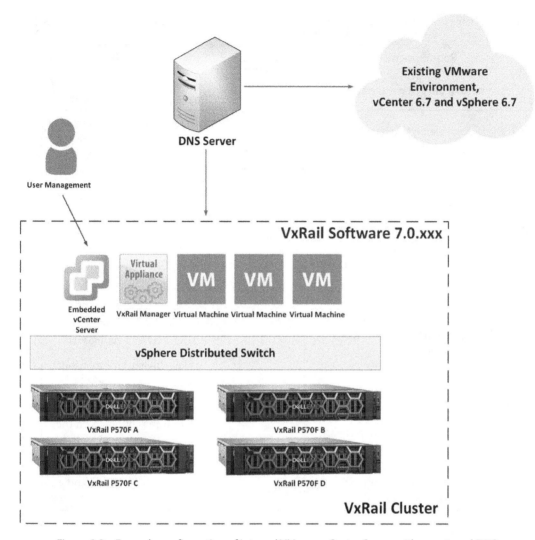

Figure 3.2 – Example configuration of internal VMware vCenter Server with an external DNS

The following table shows the compliance information for the VxRail deployment scenario in *Figure 3.2*:

Requirements	Compliance	Remark
A separate management dashboard.	Yes	The internal vCenter Server manages the VxRail cluster.
The target edition of VMware vSphere and vCenter is 7.0.	Yes	VxRail software 7.0.xxx includes VMware vSphere 7.0 and vCenter Server 7.0.
The optional VMware vCenter Server license is not required.	Yes	The VMware vCenter Server Standard license is bundled on the VxRail platform.
It allows connection to the existing DNS.	Yes	N/A.
The hardware and software support the lifecycle management of HCI.	Yes	VxRail's one-click upgrade supports this feature.
Extending into either a disaster recovery solution or an active-active solution is supported.	Yes	For a disaster recovery solution, you can enable VMware SRM for VMware vSphere. For an active-active solution, you can enable a vSAN Stretched Cluster.
It is supported by the vSphere Network Control.	Yes	This feature is bundled on the **vSphere Distributed Switch (VDS)**.
The deduplication and compression features are supported.	Yes	vSAN Enterprise license and all disks need SSDs.
It can migrate the workloads into this VxRail cluster from the existing VMware environment.	Yes	You can perform Advanced Cross vCenter vMotion.

Table 3.1 – The compliance table for this scenario

With the preceding information, you got an overview of the architecture of this VxRail deployment with the internal vCenter Server and an external DNS. This VxRail deployment is suitable for commonly used cases, for example, headquarters or active-active solutions. The standard deployment option is VxRail deployment with the internal vCenter Server and an external DNS.

The next section will discuss the other VxRail deployment option, the internal vCenter Server with an internal DNS.

> **Important Note**
> The VMware vCenter Server Standard license installed on embedded vCenter Server cannot be transferred to the external VMware vCenter Server either outside or inside the VxRail cluster.

Internal vCenter Server with an internal DNS

This section will discuss the VxRail deployment with the internal vCenter Server and an internal DNS. We'll also go through a scenario for this VxRail deployment.

Overview

In *Figure 3.3*, you can see two clusters in this VMware environment: **VxRail Cluster** and **vSphere Cluster**. The VxRail cluster is connected to an internal DNS, and the vSphere cluster is connected to an external DNS. The vSphere cluster is a standard cluster that is managed with vCenter Server. This VxRail configuration is an optional VxRail deployment option, where VxRail is installed at special sites, for example, **Demilitarized Zone** (**DMZ**) or dark sites. This deployment can simplify VxRail implementation in any environment; it does not use the existing DNS for deployment. VxRail Manager and the internal vCenter Server (embedded vCenter Server) are installed on the VxRail cluster. The host records of each VxRail node and all virtual machines (VxRail Manager, vCenter Server, and the other virtual machines) are configured in an internal DNS server that is installed on VxRail Manager during VxRail initialization. All the virtual network port groups are configured on VDS in the VxRail cluster.

If you choose the VxRail deployment with the internal vCenter Server and an internal DNS, you need to consider the following:

- Internal vCenter Server only supports managing VxRail clusters; non-VxRail clusters are not supported.

- The internal DNS is only supported when you deploy the internal vCenter Server.

- The VMware vCenter Server Standard edition license is bundled on the internal vCenter Server.

- In VxRail 7.0.xxx, the internal VMware vCenter Server is deployed with an embedded VMware PSC.

- This VxRail deployment is used in a special environment, dark site, or DMZ site.

- After this deployment is completed, the internal DNS can be migrated to an external DNS.

Figure 3.3 shows an example internal VMware vCenter Server environment with an internal DNS:

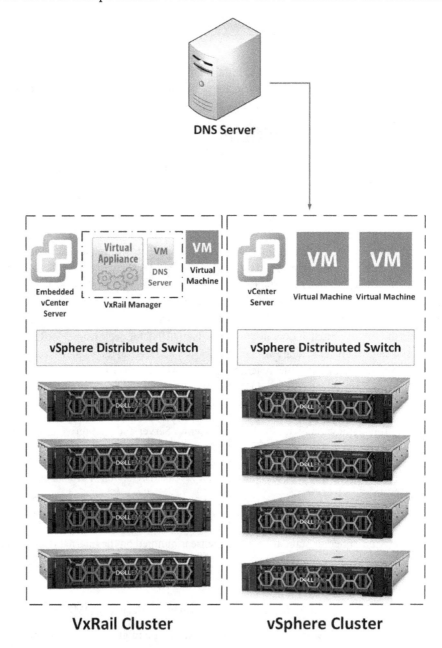

Figure 3.3 – The architecture of the internal VMware vCenter Server with an internal DNS

Now, we will discuss a scenario for this type of deployment.

Scenario

In this scenario, the customer plans to deploy a new HCI platform in their VMware environment. In the existing VMware environment, the running platform is VMware vCenter 6.7 and vSphere 6.7. They also have the following requirements for this deployment:

- The VxRail cluster is separated from the existing VMware environment. The management panel of the VxRail cluster and non-VxRail cluster must be separated.

- The target version of VMware vCenter and vSphere is 7.0 in the VxRail cluster.

- The customer does not provide the optional VMware vCenter Server license for this VxRail deployment.

- Using the existing DNS server for this VxRail deployment is not allowed.

- The lifecycle management must be supported for hardware and software components in this HCI platform.

- The HCI platform supports the vSphere Network I/O Control feature.

- The HCI platform supports deduplication and compression features.

- It supports migrating workloads into this VxRail cluster from the existing VMware environment.

- It also supports migrating host records from the internal DNS into an external DNS.

In *Figure 3.4*, you can see there are two environments; one is VMware vSphere 6.7, and the other is the VxRail cluster. The VxRail cluster is deployed with four VxRail P570F nodes. Both environments are shared and have a different DNS:

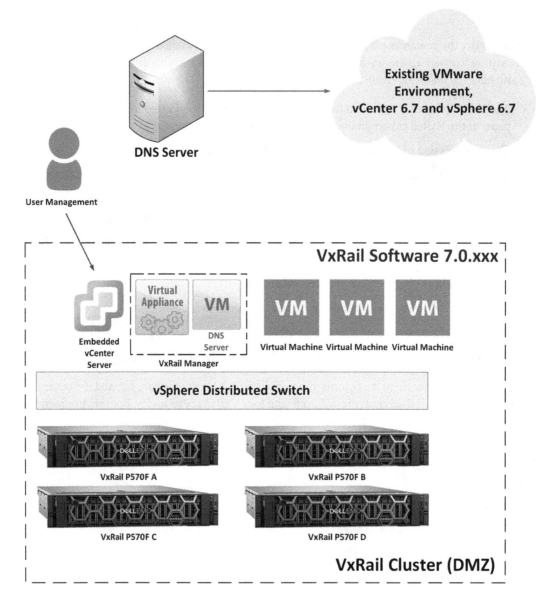

Figure 3.4 – The example configuration of the internal VMware vCenter Server with an internal DNS

The following table shows the compliance information for the VxRail deployment scenario in *Figure 3.4*:

Requirements	Compliance	Remark
A separate management dashboard.	Yes	Internal vCenter Server manages the VxRail cluster.
The target edition of VMware vSphere and vCenter is 7.0.	Yes	VxRail software 7.0.xxx includes VMware vSphere 7.0 and vCenter Server 7.0.
The optional VMware vCenter Server license is not required.	Yes	The VMware vCenter Server Standard license is bundled on the VxRail platform.
The existing DNS does not allow use for deployment.	Yes	It supports the use of the internal DNS for deployment.
The lifecycle management is supported by the hardware and software of HCI.	Yes	VxRail's one-click upgrade supports this feature.
It is supported by vSphere Network I/O Control.	Yes	This feature is bundled on the VDS.
The deduplication and compression features are supported.	Yes	These features include VMware vSAN Enterprise edition.
It can migrate workloads into this VxRail cluster from the existing VMware environment.	Yes	You can perform Advanced Cross vCenter vMotion.
The host records can be migrated from the internal DNS into an external DNS.	Yes	N/A.

Table 3.2 – The compliance table of this scenario

With the preceding information, you got an overview of the architecture of this VxRail deployment with the internal vCenter Server and an internal DNS. This VxRail deployment is suitable for special environments, for example, DMZ and dark sites. An optional deployment option is a VxRail deployment with the internal vCenter Server and an internal DNS. The next section will discuss the other VxRail deployment option, the external vCenter Server with an external DNS.

External vCenter Server with an external DNS

This section will discuss a VxRail deployment with the external vCenter Server and an external DNS, and run through a scenario for this VxRail deployment.

Overview

In *Figure 3.5*, you can see two clusters connected to an external DNS in this VMware environment: **VxRail Cluster** and **vSphere Cluster**. Both clusters are managed with the external vCenter Server. The external vCenter Server could be a new vCenter Server instance or a vCenter Server instance in the existing VMware infrastructure environment. If you choose this VxRail deployment, you can manage and monitor the VxRail cluster and the existing VMware infrastructure environment in a single management interface, with VxRail Manager installed on the VxRail cluster. Each VxRail node and all virtual machines (VxRail Manager and the other virtual machines) are configured in the external DNS server's host records. All the virtual network port groups are configured on VDS in the VxRail cluster.

If you choose VxRail deployment with the external vCenter Server and an external DNS, you need to consider the following:

- External vCenter Server supports managing both VxRail clusters and non-VxRail clusters in a single management dashboard.

- External vCenter supports VxRail standard deployment and vSAN Stretched Clusters; vSAN two-node cluster deployment is not supported.

- The VMware vCenter Server Standard edition license is not bundled on the external vCenter Server; you need to prepare the optional vCenter Server Standard license for the external vCenter Server.

- The external DNS server is required when using the external vCenter Server with VxRail.

- The external vCenter Server supports the management of multiple VxRail clusters.

Figure 3.5 shows an example environment of the external VMware vCenter Server with an external DNS:

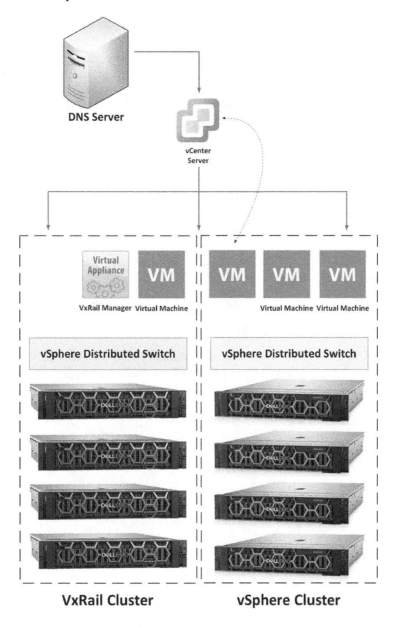

Figure 3.5 – The architecture of the external VMware vCenter Server with an external DNS

Now we will discuss a scenario for this type of deployment.

Scenario

In this scenario, the customer plans to deploy a new HCI platform in their VMware environment. In the existing VMware environment, the running platform is VMware vCenter 7.0 and vSphere 6.7. They expect that this HCI and existing VMware infrastructure can be managed and monitored in a single management interface. They also have the following requirements for this deployment:

- The VxRail cluster and non-VxRail cluster can be managed and monitored with the existing vCenter Server instance.

- The target version of VMware vCenter and vSphere is 7.0 in the VxRail cluster.

- They are using the existing DNS and AD server for this VxRail deployment.

- Lifecycle management must be supported for hardware and software components in this HCI platform.

- The HCI platform supports extending the VxRail cluster into the disaster recovery solution or active-active solution.

- The HCI platform supports the vSphere Network I/O Control feature.

- The HCI platform supports deduplication and compression features.

- Migrating workloads into this VxRail cluster from the existing VMware environment is supported.

In *Figure 3.6*, you can see there are two environments: one is VMware vSphere 6.7 and the other is the VxRail cluster. The VxRail cluster is deployed with four VxRail P570F nodes. Both environments share the external DNS:

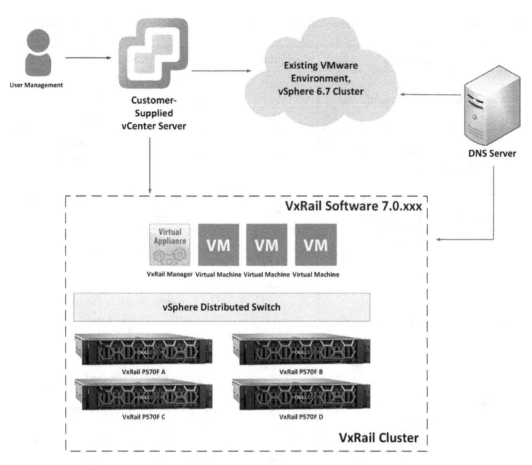

Figure 3.6 – The example configuration of the external VMware vCenter Server with an external DNS

The following table shows the compliance information for the VxRail deployment scenario in *Figure 3.6*:

Requirements	Compliance	Remark
A single management dashboard.	Yes	The external vCenter Server supports managing and monitoring VxRail and non-VxRail systems.
The target edition of VMware vSphere and vCenter is 7.0.	Yes	VxRail software 7.0.xxx includes VMware vSphere 7.0 and vCenter Server 7.0.
Using the existing DNS and AD server for this configuration.	Yes	It supports the use of an external DNS for deployment.
The lifecycle management is supported by the hardware and software of HCI.	Yes	VxRail's one-click upgrade supports this feature.
Extending the VxRail cluster into either a disaster recovery solution or an active-active solution is supported.	Yes	For a disaster recovery solution, you can enable VMware SRM or VMware vSphere. For an active-active solution, you can enable a vSAN Stretched Cluster.
It is supported by vSphere Network I/O Control.	Yes	This feature is bundled on the VDS.
Deduplication and compression features are supported.	Yes	These features include VMware vSAN Enterprise edition and require all SSDs in each node.
It can migrate workloads into this VxRail cluster from the existing VMware environment.	Yes	You can perform Advanced Cross vCenter vMotion.

Table 3.3 – The compliance table of this scenario

With the preceding information, you got an overview of the architecture of this VxRail deployment with the external vCenter Server and an external DNS. This VxRail deployment is suitable for commonly used cases, for example, headquarters or active-active solutions. This configuration option is a VxRail deployment with the external vCenter Server and an external DNS.

The next section will discuss the other VxRail deployment option, the external vCenter Server with a customer-supplied VDS.

> **Important Note**
> Please ensure the external VMware vCenter Server version is listed in the Dell KB 520355 when using VxRail with the external VMware vCenter Server.

External vCenter Server with customer-supplied VDS

This section discusses a VxRail deployment with the external vCenter Server and a customer-supplied VDS and goes through a scenario for this VxRail deployment.

Overview

In *Figure 3.7*, you can see three VxRail clusters in this VMware environment: VxRail clusters **A**, **B**, and **C**. All clusters are managed with the external vCenter Server. The external vCenter Server can be a new vCenter Server instance or a vCenter Server instance in the existing VMware infrastructure environment. Starting from VxRail 7.0.010, a deployment option is available where a single customer-supplied VDS is created via the existing vCenter Server instance. If you choose this VxRail deployment, you can manage and monitor each VxRail cluster in a single management interface and the customer-supplied VDS spans across each VxRail cluster. VxRail Manager is installed on the VxRail cluster. Each VxRail node and all virtual machines (VxRail Manager and the other virtual machines) are configured in the external DNS server's host records.

If you choose VxRail deployment with the external vCenter Server with a customer-supplied VDS, you need to consider the following:

- The customer-supplied VDS can only be created via a **JavaScript Object Notation (JSON)** file configured through the VxRail deployment wizard. You must set up the external VDS and its network port group prior to connecting the VxRail cluster.

- It manages all VxRail VDSs across multiple VxRail clusters via a single vCenter Server instance.

- This deployment simplifies VxRail daily and maintenance operations.

- The VMware vCenter Server Standard edition license is not bundled on the external vCenter Server; you need to prepare the optional vCenter Server Standard license for the external vCenter Server.

- An external DNS server is required when using an external vCenter Server with VxRail.

- External vCenter Server supports the management of multiple VxRail clusters.

Figure 3.7 shows an example environment of the external VMware vCenter Server with a customer-supplied VDS:

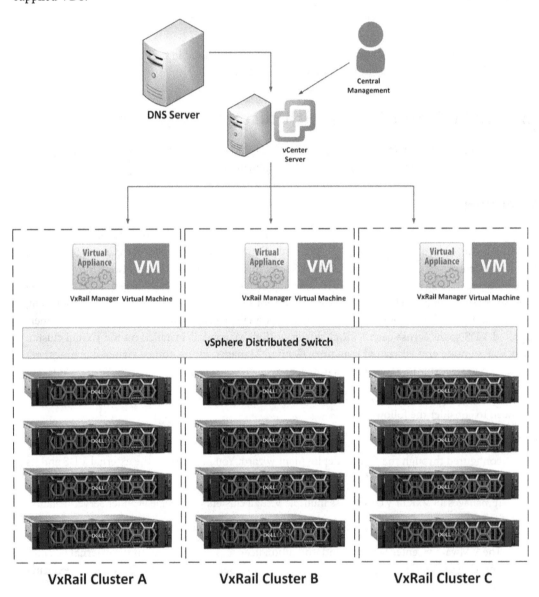

Figure 3.7 – The architecture of external VMware vCenter Server with a customer-supplied VDS

Now, we will discuss a scenario for this type of deployment.

Scenario

In this scenario, the customer plans to deploy a new HCI platform in their VMware environment. In the existing VMware environment, there is a VxRail cluster running that is managed with the external vCenter Server. They expect the new HCI and existing HCI platform to be managed and monitored in a single management interface. They also have the following requirements for this deployment:

- The VxRail cluster and non-VxRail cluster can be managed and monitored with the existing vCenter Server instance.

- The target version of VMware vCenter and vSphere is 7.0 in the VxRail cluster.

- They are using the existing DNS and AD server for this VxRail deployment.

- They want to manage all VxRail VDSs across multiple clusters under a single VDS via a single vCenter Server instance.

- It can simplify VxRail daily and maintenance operations.

- Lifecycle management must be supported for hardware and software components in this HCI platform.

- The HCI platform supports the vSphere Network I/O Control feature.

- The HCI platform supports deduplication and compression features.

- Migrating workloads into this VxRail cluster from the existing VMware environment is supported.

In *Figure 3.8*, you can see there are two VxRail clusters: **VxRail Cluster A** (VxRail software 7.0.130) and **VxRail Cluster B** (VxRail software 7.0.320). The user can manage and monitor both VxRail clusters through the external vCenter Server. **VxRail Cluster A** is deployed with four VxRail P570F nodes, and **VxRail Cluster B** is deployed with four VxRail P570 nodes. Both VxRail clusters share the existing DNS and all VxRail VDSs across multiple clusters are under a single VDS via a single vCenter Server instance:

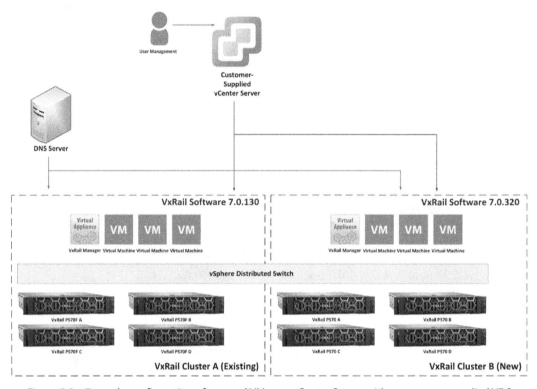

Figure 3.8 – Example configuration of external VMware vCenter Server with a customer-supplied VDS

The following table shows the compliance information for the VxRail deployment scenario in *Figure 3.8*:

Requirements	Compliance	Remark
A single management dashboard.	Yes	The external vCenter Server supports managing and monitoring multiple VxRail clusters.
The target edition of VMware vSphere and vCenter is 7.0.	Yes	VxRail software 7.0.xxx includes VMware vSphere 7.0 and vCenter Server 7.0.
Uses the existing DNS and AD server for this configuration.	Yes	It supports the use of an external DNS for deployment.
Manage all VxRail VDSs across multiple clusters under a single VDS via a single vCenter Server instance.	Yes	Starting from VxRail 7.0.010, this feature is available.
It can simplify VxRail daily and maintenance operations.	Yes	All operational tasks can be executed in vCenter Server via the VxRail Manager plugin for vCenter.
Lifecycle management is supported by the hardware and software of HCI.	Yes	VxRail's one-click upgrade supports this feature.
It is supported by vSphere Network I/O Control.	Yes	This feature is bundled on the VDS.
Deduplication and compression features are supported.	Yes	These features include VMware vSAN Enterprise edition and require all SSDs.
It can migrate workloads into this VxRail cluster from the existing VMware environment.	Yes	You can perform Advanced Cross vCenter vMotion.

Table 3.4 – The compliance table of this scenario

With the preceding information, you got an overview of the architecture of this VxRail deployment with the external vCenter Server with a customer-supplied VDS. This VxRail deployment is suitable for the management of multiple VxRail clusters in a single customer-supplied VDS.

> **Important Note**
>
> For more information about customer-supplied VDS deployment, you can use this link:
> https://www.delltechnologies.com/asset/en-us/products/converged-infrastructure/technical-support/h15300-vxrail-network-guide.pdf.

Summary

In this chapter, you learned about the design of vCenter Server for the VxRail 7 system, including the internal vCenter Server with an external DNS and an internal DNS, the external vCenter Server with an external DNS, and the internal vCenter Server with a customer-supplied VDS. The following table shows a summary of the design of each vCenter Server type:

VMware vCenter Server	Cluster type	Domain Name Server	Deployed internal vCenter Server to VxRail cluster	Deployed external vCenter Server to VxRail cluster
Internal	Standard	External	Supported	Not supported
	vSAN Stretched	External	Supported	Not supported
	vSAN two-node	External	Not supported	Not supported
Internal	Standard	Internal	Supported	Not supported
	vSAN Stretched	Internal	Supported	Not supported
	vSAN two-node	Internal	Not supported	Not supported
External	Standard	External	Supported	Supported
	vSAN Stretched	External	Supported	Supported
	vSAN two-node	External	Not supported	Default
External with customer-supplied VDS	Standard	External	Supported	Supported
	vSAN Stretched	External	Supported	Supported
	vSAN two-node	External	Not supported	Supported

Table 3.5 – A summary of each vCenter Server design for VxRail deployment

You will learn about the design of vSAN storage policies on the VxRail Appliance system in the next chapter, including an overview of vSAN objects, components, and vSAN storage policies.

Questions

The following is a short list of review questions to help reinforce your learning and help you identify areas that require some improvement:

1. Which VxRail deployment options are supported?

    A. Internal vCenter Server with an external DNS

    B. Internal vCenter Server with an internal DNS

    C. External vCenter Server with an external DNS

    D. External vCenter Server with a customer-supplied VDS

    E. All of the above

2. Which statement is incorrect regarding VxRail deployment with the internal vCenter Server and an external DNS?

    A. The VMware vCenter Server license is bundled on the internal vCenter Server.

    B. The internal vCenter Server supports the management of multiple VxRail clusters.

    C. The internal vCenter Server supports the management of vSAN two-node clusters.

    D. In VxRail software 7.0.xxx, the internal VMware vCenter Server is deployed with an embedded VMware PSC.

    E. All of the above.

3. Which VxRail deployment options do not require the optional vCenter Server license?

    A. Internal vCenter Server with an external DNS

    B. Internal vCenter Server with an internal DNS

    C. External vCenter Server with an external DNS

    D. External vCenter Server with a customer-supplied VDS

    E. All of the above

4. Which scenario is used for VxRail deployment with the internal vCenter Server and an internal DNS?

    A. Active-passive data center solution

    B. Active-active data center solution

    C. vSAN two-node cluster

    D. DMZ environment

    E. Headquarter data center

5. Which statement is incorrect regarding VxRail deployment with the internal vCenter Server and an internal DNS?

    A. This kind of VxRail deployment can be used for all common cases.

    B. The DNS server is installed on VxRail Manager.

    C. Non-VxRail clusters do not support being managed by the internal vCenter Server.

    D. This kind of VxRail deployment can be used for dark sites or DMZ sites.

    E. The host records can be migrated from the internal DNS into the external DNS.

    F. None of the above

6. Which statements are incorrect regarding VxRail deployment with the external vCenter Server and an external DNS?

    A. This kind of VxRail deployment can be used for all common cases.

    B. This kind of VxRail deployment is only supported with an external DNS.

    C. Non-VxRail clusters do not support being managed by the external vCenter Server.

    D. You do not need to provide the optional VMware vCenter Server license.

    E. You need to provide the optional VMware vCenter Server license.

    F. None of the above.

7. Which VxRail deployments are not supported with vSAN Stretched Clusters?

    A. Internal vCenter Server with an external DNS (VxRail 4.7.xxx)

    B. Internal vCenter Server with an internal DNS (VxRail 4.7.xxx)

    C. External vCenter Server with an external DNS

    D. External vCenter Server with a customer-supplied VDS

    E. All of the above

8. Which VxRail deployments are not supported with the vSAN two-node cluster?

    A. Internal vCenter Server with an external DNS

    B. Internal vCenter Server with an internal DNS

    C. External vCenter Server with an external DNS

    D. External vCenter Server with a customer-supplied VDS

    E. All of the above

9. If the external vCenter Server version is 7.0 U2 (7.0.2), what are the supported VxRail release versions for this deployment?

    A. VxRail 7.0.100

    B. VxRail 7.0.130

    C. VxRail 7.0.200

    D. VxRail 7.0.300

    E. VxRail 4.7.540

    F. All of the above

10. Which VxRail software editions support a customer-supplied VDS?

    A. VxRail 4.7.xxx

    B. VxRail 7.0.000

    C. VxRail 7.0.010

    D. VxRail 7.0.130

    E. VxRail 7.0.200 or above

    F. All of the above

11. Which VxRail software editions support link aggregation on a customer-supplied VDS?

    A. VxRail 4.7.xxx

    B. VxRail 7.0.000

    C. VxRail 7.0.010

    D. VxRail 7.0.130

    E. VxRail 7.0.200 or above

    F. All of the above

12. Which VMware vCenter Server deployment can be supported with the management of all VxRail VDSs across multiple VxRail clusters via a single vCenter Server instance?

    A. Internal vCenter Server with an external DNS

    B. Internal vCenter Server with an internal DNS

    C. External vCenter Server with an external DNS

    D. External vCenter Server with a customer-supplied VDS

    E. All of the above

4
Design of vSAN Storage Policies

In the previous chapter, you learned about the design of vCenter Server for the VxRail 7.x system, including an internal vCenter Server instance with an external DNS and an internal DNS, an external vCenter Server instance with an external DNS, and an internal vCenter Server instance with a customer-supplied virtual distributed switch. Now, you understand the advantages and disadvantages of each kind of VxRail deployment.

VMware vSAN is a key component of the VxRail system. It can automatically build up a local vSAN datastore across the VxRail cluster during VxRail initialization. VMware vSAN can provide different features based on your choice of vSAN license editions on the VxRail cluster – for example, Standard edition, Advanced edition, and Enterprise edition. You can non-disruptively expand the capacity and performance by adding nodes to a VxRail cluster (scaling out) or increase the capacity by adding disks to each node (scaling up). The vSAN storage policy is a good tool that can deliver different storage requirements on the VxRail 7.x system and it can also deliver the data recovery services for each **virtual machine (VM)**.

This chapter will provide an overview of VMware vSAN on the VxRail 7.x system, including the vSAN architecture, vSAN components, vSAN disk groups, vSAN storage policies, and vSAN HCI Mesh.

This chapter includes the following topics:

- An overview of VMware vSAN on VxRail
- An overview of vSAN objects and components
- VMware vSAN storage policies
- VMware vSAN HCI Mesh

An overview of VMware vSAN on VxRail

VMware vSAN is a **Software-Defined Storage** (SDS) solution that is used in collaboration with the VMware vSphere hypervisor. The VMware vSAN solution can provide shared storage for VMs. In *Figure 4.1*, vSAN is a vSphere cluster feature that virtualizes the local physical storage in the pool of the **Hard Disk Drive** (HDD) and **Solid-State Drive** (SSD) on the ESXi hosts in a cluster, building them into a unified data store. Each ESXi host requires a minimum disk group that includes a SSD and HDD. vSAN can provide enterprise storage that is powerful, flexible, and user-friendly. This solution can be deployed in different environments – for example, the core, the edge, and the cloud. When you build the VxRail cluster on day one, the vSAN datastore is created automatically.

The vSAN datastore includes the following characteristics:

- All nodes in the VxRail cluster must be connected to a vSAN Layer 2 or Layer 3 network. It supports a maximum of a 1 **millisecond (ms) Round-Trip Time (RTT)** for VxRail standard clusters between all nodes in the cluster.

- Using a **vSphere Distributed Switch** (vDS) and **Network IO Control** (NIOC) is recommended.

- VMware vSAN is a single vSAN datastore accessible to all nodes in the VxRail cluster.

- You can create different service levels for each VM or **Virtual Machine Disk** (VMDK) per VM using vSAN storage policies.

- vSAN supports Hybrid mode and All-Flash mode on the VxRail cluster.

- The Hybrid and All-Flash **disk groups** cannot mix in the same VxRail cluster.

- vSAN does not support **Raw Device Mapping** (RDM).

- A vSAN datastore can be supported for mounting by a maximum of five vSAN client clusters.

- vSAN Stretched Clusters and two-node configurations are not supported in combination with HCI Mesh.

In *Figure 4.1*, there is a VxRail cluster with four nodes in this environment and one disk group installed on each VxRail, including one SSD for the cache tier and two capacity disks for the capacity tier. The vSAN datastore's capacity depends on the number of capacity disks per VxRail node and the number of VxRail nodes in the cluster. For example, if the VxRail cluster includes four nodes and each node installs two 2 TB capacity disks, the approximate storage capacity is 16 TB (4 x 4 TB). vSAN supports two configurations – Hybrid and All-Flash mode:

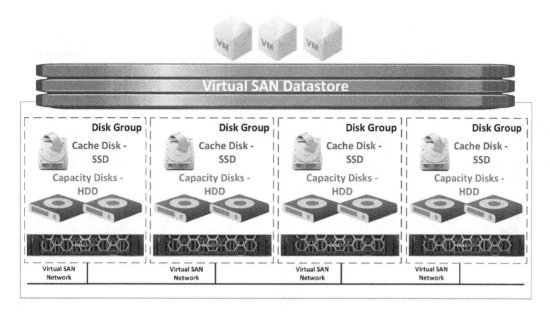

Figure 4.1 – The architecture of VMware vSAN on the VxRail 7.x system

> **Important Note**
>
> In the initial deployment, the first three VxRail nodes in a cluster must be identical models. VxRail Hybrid and All-Flash nodes cannot mix in a VxRail cluster. VxRail S Series does not support the All-Flash model.

In *Figure 4.2*, you can configure a disk group with either Hybrid or All-Flash configurations in the VxRail cluster. Each VxRail model can support a maximum of four disk groups, each of which must have one flash device and one or more capacity devices (up to five devices). The number of supported disk groups depends on the VxRail model; for example, VxRail E Series supports a maximum of two disk groups, or VxRail E Series supports a maximum of four disk groups. In the Hybrid disk group, 70% of the available cache is used for frequently-read data, and 30% of the available cache is used for write buffering. In the All-Flash disk group, the flash devices are used in a two-tier format for caching and capacity, and 100% of the available cache is used for write buffering:

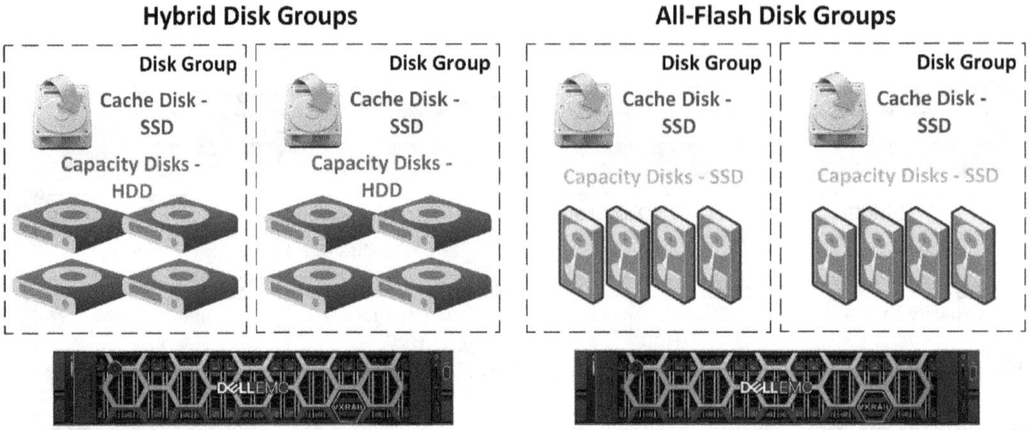

Hybrid Disk Groups

Disk Group	Disk Group
Cache Disk - SSD	Cache Disk - SSD
Capacity Disks - HDD	Capacity Disks - HDD

All-Flash Disk Groups

Disk Group	Disk Group
Cache Disk - SSD	Cache Disk - SSD
Capacity Disks - SSD	Capacity Disks - SSD

Figure 4.2 – The difference between the Hybrid disk group and the All-Flash disk group

> **Important Note**
> The SSD and NVMe devices support the cache tier on the VxRail cluster. The SSD, NVMe, SAS, SATA, and vSAS devices support the capacity tier on the VxRail cluster.

In a VMware vSAN 7.x cluster, you can also configure a disk group with either Hybrid or All-Flash configurations manually, but each vSAN node can support a maximum of *five disk groups*, each of which must have one flash device and one or more capacity devices (**up to seven devices**). Compared to the VxRail cluster, the maximum disk group is different from the VMware vSAN cluster. A VxRail node can only support up to four disk groups and the capacity tier supports up to six capacity devices. In each VxRail model, the disk group configuration rules are specific to vSAN configurations. For the disk group upgrades on each VxRail Series, we will discuss the details in *Chapter 5, Design of Cluster Expansion*.

The next section discusses the overview of vSAN objects and components.

An overview of vSAN objects and components

When you create VMs in a VxRail cluster, you must specify the vSAN storage policy (we will discuss the details in the next section) for each VM. It will automatically create each object's vSAN component (that is, the VMDK) and allocate it to each VxRail node when you apply the vSAN storage policy to the VM. vSAN manages and stores data called objects. Each object on the vSAN datastore includes data, metadata, and a unique ID. The most common vSAN objects are the VDMKs. In traditional storage, you need to define the storage pools, the RAID level protection, and the hot spare. Then, you can create VMs or build your application into this storage. All storage configurations must be completed on day one. The vSAN architecture is different from traditional storage. vSAN objects are made of components determined by vSAN storage policies. You can change or update the **Failure Tolerance Method** (FTM) and **Failures To Tolerate** (FTT) of the VM at any time if the storage resource requirement is fulfilled in the VxRail cluster. In an example in *Figure 4.3*, there is a VM allocated to the VxRail cluster with four nodes and it is assigned the vSAN storage policy with **FTM = RAID-1** and **FTT = 1** in this VM. The vSAN object (the VMDK) included has three components, two replicas, and one witness. Each component is allocated to each VxRail node. Each vSAN object has the following state:

- `Active`: The vSAN object is accessible.

- `Absent`: The vSAN object is inaccessible, but no explicit error codes were detected.

- `Degraded`: The vSAN object is inaccessible with explicit error codes detected.

- `Active-Stale`: The sequence numbers are not consistent.

The level of FTT defines the level of resilience. The FTM defines the data layout approach. In this example, the fault tolerance method is set to RAID-1 mirroring, and FTT are set to 1:

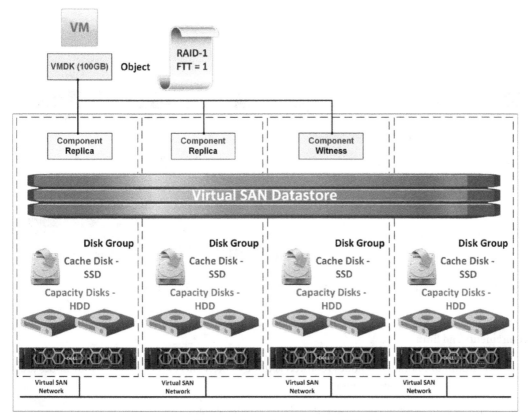

Figure 4.3 – The sample configuration of vSAN objects

Now, we will discuss the architecture of different FTMs in the next section, including RAID-1, RAID-5, and RAID-6.

Failures to tolerate with RAID-1 mirroring

This section will discuss an overview of FTMs with RAID-1 on the VxRail cluster. In *Figure 4.4*, there is a VxRail cluster with five nodes (**VxRail Node A, B, C, D**, and **E**) in this scenario. And one VM with one 100 GB VMDK configured with **RAID-1** and **FTT = 1**. There are three vSAN components: two **Replica** components and one **Witness** component. Each vSAN component is allocated to each node. Applying this vSAN storage policy to the VM will mirror the **Replica** component (the 100 GB VMDK) from **VxRail Node A** into **VxRail Node B**. The **Witness** component is used to determine a quorum:

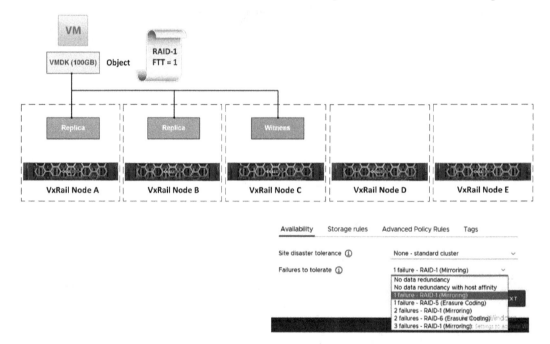

Figure 4.4 – The diagram of vSAN objects in RAID-1 and FTT = 1

If you choose the vSAN policy with **RAID-1** and **FTT = 1**, you need to consider the following requirements and limitations:

- The minimum number of nodes required is three in the same VxRail cluster – four nodes are the recommended configuration.
- It supports the VxRail Hybrid and All-Flash configuration.
- It supports vSAN Stretched Clusters.

- It supports the VMware vSAN standard, advanced, enterprise, and enterprise plus editions.

- It allows one node failure in the VxRail cluster if FTT is set to one – four nodes is the recommended configuration.

- It allows two-node failure in the VxRail cluster if FTT is set to two – six nodes is the recommended configuration.

- It allows three-node failure in the VxRail cluster if FTT is set to three; eight nodes is the recommended configuration.

According to this scenario, you have learned the architecture of FTMs with RAID-1 on the VxRail cluster. We will discuss the RAID-5 erasure coding scenario in the next section.

Failures to tolerate with RAID-5 erasure coding

This section will discuss the overview of FTMs with RAID-5 on the VxRail cluster. In *Figure 4.5*, there is a VxRail cluster with five nodes (**VxRail Node A, B, C, D, and E**), and one VM with one 100 GB VMDK configured with **RAID-5** and **FTT = 1**. There are four vSAN components, three replicas, and one **Parity** component. Each vSAN component is allocated to each node when you apply this vSAN storage policy to the VM – after then, it separates into three replica components (100 GB VMDK) and one parity of the VM. Each component is stored across four hosts (**VxRail Node A, B, C, and D**):

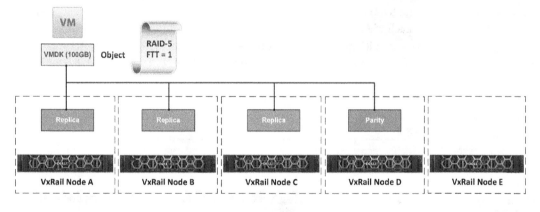

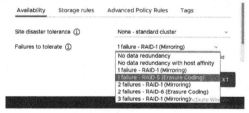

Figure 4.5 – The diagram of vSAN objects in RAID-5 and FTT = 1

If you choose the vSAN policy with **RAID-5** and **FTT = 1**, you need to consider the following requirements and limitations:

- The minimum number of nodes is four in the same VxRail cluster, and five nodes are the recommended configuration.
- It only supports the VxRail All-Flash configuration.
- It supports vSAN Stretched Clusters.
- It supports the VMware vSAN advanced, enterprise, and enterprise plus editions.
- It allows one node failure in the VxRail cluster if FTT is set to one – five nodes are the recommended configuration.
- A space reduction of 30% is guaranteed with RAID-5.

With this scenario, you have learned about the architecture of FTMs with RAID-5 on the VxRail cluster. We will discuss the RAID-6 erasure coding scenario in the next section.

Failures to tolerate with RAID-6 erasure coding

This section will discuss the overview of FTMs with RAID-6 on the VxRail cluster. In *Figure 4.6*, there is a VxRail cluster with seven nodes (**VxRail Node A**, **B**, **C**, **D**, **E**, **F**, and **G**) and one VM with one 100 GB VMDK configured with **RAID-6** and **FTT = 2**. There are six vSAN components, four replicas, and two **Parity** components. Each vSAN component is allocated to each node when you apply this vSAN storage policy to the VM – then it separates into four replica components (100 GB VMDK) and two Parity components. Each component is stored across six hosts (**VxRail Node A**, **B**, **C**, **D**, **E**, and **F**).

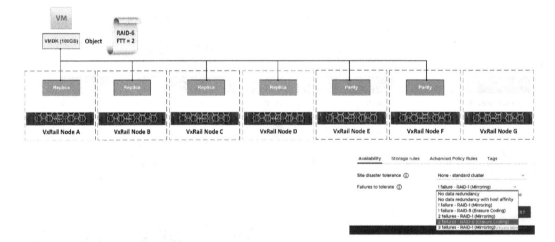

Figure 4.6 – The diagram of vSAN objects in RAID-6 and FTT = 2

If you choose the vSAN policy with **RAID-6** and **FTT = 2,** you need to consider the following requirements and limitations:

- The minimum number of nodes is six in the same VxRail cluster, and seven nodes are the recommended configuration.

- It only supports the VxRail All-Flash configuration.

- It supports vSAN Stretched Clusters.

- It supports the VMware vSAN advanced, enterprise, and enterprise plus editions.

- It allows two-node failure in the VxRail cluster if FTT is set to two; seven nodes are the recommended configuration.

- It guarantees a space reduction of 50% with RAID-6.

Important Note

Both RAID-5 and RAID-6 are only supported by the All-Flash disk group in the VxRail cluster. And the vSAN license edition must be Advanced, Enterprise, or Enterprise Plus.

With this scenario, you have learned about the architecture of FTMs with RAID-6 on a VxRail cluster. We will discuss vSAN storage policies in the next section.

VMware vSAN storage policies

The vSAN storage policy is used to define the VM storage requirements of performance and availability. After vSAN storage is applied to a VM, a number of vSAN component replicas and copies are created automatically based on the vSAN storage policy. vSAN supports the following RAID level protection. *Table 4.1* includes a description of each supported RAID level on vSAN:

RAID Level	Description
RAID-0	This is a striped volume and has the fastest performance but no redundancy.
RAID-1	This is a mirrored volume with good performance and full redundancy with 200% capacity usage.
RAID-10	This is a mirrored and stripped volume with the best performance and full redundancy with 200% capacity usage.
RAID-5	This is a stripped volume with parity. It gives good performance with redundancy and requires a minimum of five nodes.
RAID-6	This is a stripped volume with double parity. It has good performance with redundancy that requires a minimum of seven nodes.

Table 4.1 – Descriptions of FTMs on the vSAN storage policies

When you create the vSAN storage policy based on your storage requirements for a VM and its VMDKs, you can define the different storage rules, including encryption services, space efficiency, and storage tiers.

In *Figure 4.7*, you can go to the **Storage rules** tab in the **Create VM Storage Policy** wizard if you want to enable storage rules on your vSAN storage policies. You can select the following storage rules:

- **Encryption services** is used to define the encryption rules for the VM:

    - **Data-At-Rest encryption**: Enables encryption on the VM

    - **No encryption**: Does not enable encryption on the VM

    - **No preference**: Makes the VM compatible with both **Data-At-Rest encryption** and **No encryption**

- **Space efficiency** is used to save storage space for the VM:

    - **Deduplication and compression**: Enables both deduplication and compression on VMs. Both features are only available on the vSAN All-Flash cluster.

    - **Compression only**: Enables compression on the VM. This feature is only available on the vSAN All-Flash cluster.

    - **No space efficiency**: Does not enable the space efficiency feature on the VM.

    - **No preference**: Makes the VM compatible with all of the preceding options.

- **Storage tier** is used to define the storage tier on the VM:

    - **All Flash**: This is supported by the vSAN All-Flash environment.

    - **Hybrid**: This is supported by the vSAN Hybrid environment.

    - **No preference**: Makes the VM compatible with All-Flash and Hybrid environments.

In the **Storage rules** tab, you can define all of these features in the **Create VM Storage Policy** wizard:

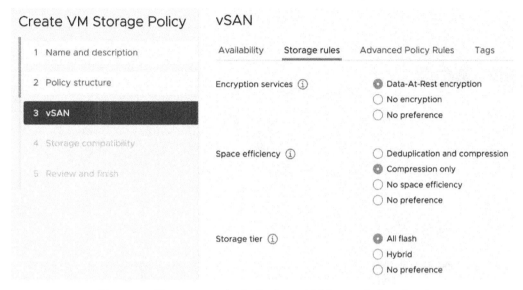

Figure 4.7 – The storage rules in the Create VM Storage Policy wizard

You can also define advanced policy rules for the VM. You can choose from the following options:

- **Number of disk stripes per object**: This defines the number of stripes for each vSAN object (or VMDK). The default setting is one; if you change it to two, the vSAN object is striped across two physical disks.

- **IOPS limit for object**: This defines the IOPS for the drive.

- **Object space reservation**: This defines the space type of the vSAN object, **Thick provisioning** or **Thin provisioning**.

- **Flash read cache reservation**: This defines the flash read cache for each vSAN object.

- **Disable object checksum**: If you enable this option, the storage object will not calculate the checksum information.

- **Force provisioning**: If you enable this option, the vSAN object can be provisioned if the vSAN storage policy cannot satisfy the resources available in the cluster.

In the **Advanced Policy Rules** tab, you can define all of the preceding features in the **Create VM Storage Policy** wizard, as shown in *Figure 4.8*:

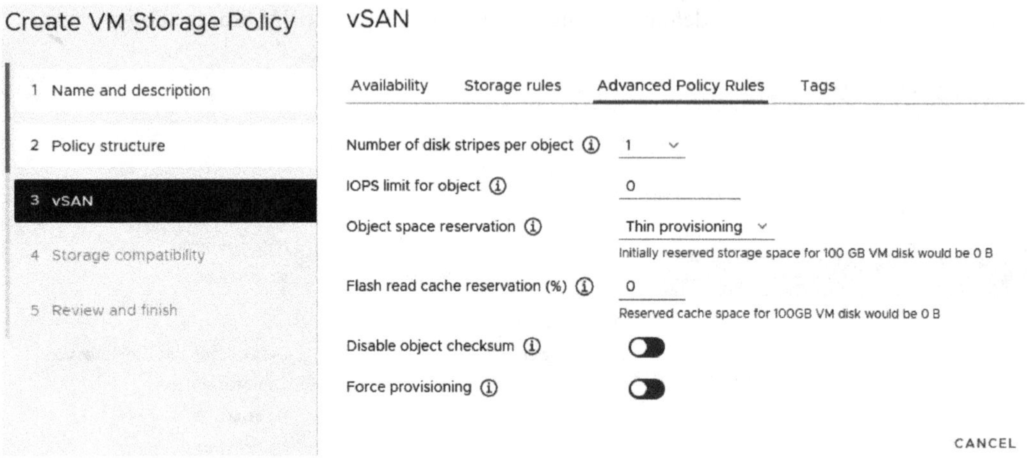

Figure 4.8 – The advanced policy rules in the Create VM Storage Policy wizard

So far, we have had an overview of the vSAN storage policy. We will discuss a scenario involving vSAN storage policies on the VxRail cluster in the next section.

vSAN storage policy scenario

This section will discuss the vSAN component's behavior on the VxRail cluster after updating the different vSAN storage policies, including RAID-1 and RAID-5. In the following scenarios, one disk group is required per node. In *Figure 4.9*, there is a VxRail cluster with five nodes and a VM with a 100 GB VMDK running on this cluster. The vSAN storage policy configures the following storage rules and applies them to the VM. There are five vSAN components, four **Replica** components (50 GB), and one **Witness** component in this scenario. In **VxRail Node A**, two components are allocated in **Disk 1** and **Disk 2**. In **VxRail Node B**, two components are allocated in **Disk 1** and **Disk 2**. The **Witness** component allocates the disk on **VxRail Node C**. The vSAN storage policy includes the following parameters:

- The **FTM** is **RAID-1.**
- **FTT** is **1.**
- **Stripe Width (SW)** is **2.**

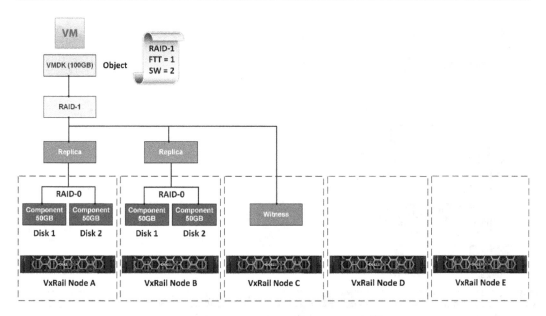

Figure 4.9 – The vSAN storage policy with FTM = RAID-1, FTT = 1, and SW = 2

Important Note

If the vSAN objects are greater than 255 GB in size, vSAN automatically divides them into multiple components.

The following is the summary of the vSAN components if you apply the vSAN storage policy with **FTM = RAID-1**, **FTT = 1**, and **SW = 2**:

- The five vSAN components are created in the VxRail cluster.
- The size of each vSAN component is 50 GB and the witness component is not included.
- One node can fail in the VxRail cluster.
- Each vSAN component is written onto each disk.

According to the scenario in *Figure 4.9*, you now understand the behavior of the vSAN components. Now, we will update the **SW** variable to **3** in the vSAN storage policy:

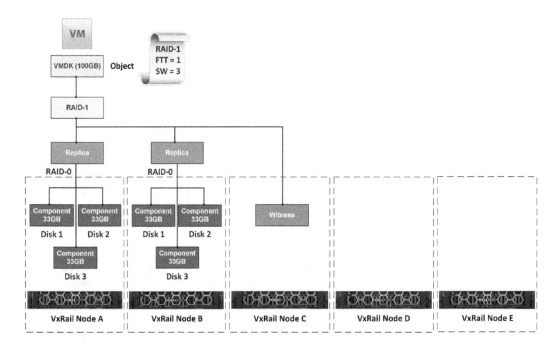

Figure 4.10 – The vSAN storage policy with FTM = RAID-1, FTT = 1, and SW = 3

The following is the summary of the vSAN components if you apply the vSAN storage policy with **FTM = RAID-1, FTT = 1**, and **SW = 3**:

- The seven vSAN components are created in the VxRail cluster.

- The size of each vSAN component is around 33 GB and the **Witness** component is not included.

- One node can fail in the VxRail cluster.

- Each vSAN component is written onto each disk.

According to the scenario in *Figure 4.10*, you now understand the number of vSAN components will be changed to seven because the **SW** parameter changes to **3** in the vSAN storage policy. Now, we will discuss another scenario.

In *Figure 4.11*, there are eight vSAN components if you **FTM = RAID-5** and **FTT = 2**, including six **Replica** components (16.5 GB) and two **Parity** components (16.5 GB):

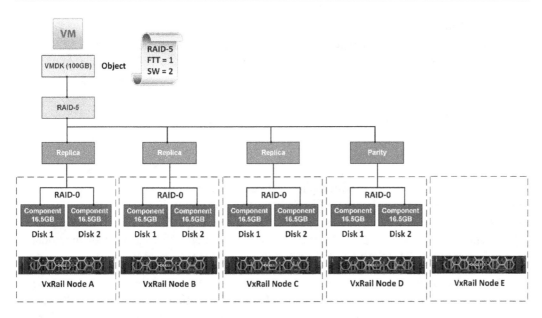

Figure 4.11 – The vSAN storage policy with FTM = RAID-5, FTT = 1, and SW = 2

The following is the summary of the vSAN components if you apply the vSAN storage policy with **FTM = RAID-5**, **FTT = 1**, and **SW = 2**:

- The eight vSAN components are created in the VxRail cluster.

- The size of each vSAN component is around 16.5 GB and there is no witness component in RAID-5.

- One node can fail in the VxRail cluster.

- Each vSAN component is written onto each disk.

According to the preceding scenarios, you have learned about the behavior of the vSAN components in the vSAN storage policies with RAID-1 and RAID-5.

Table 4.2 is a summary of FTT and FTMs on the vSAN storage policy:

Failures to Tolerate	RAID-1 (Mirroring)		RAID-5/6 (Erasure Coding)		RAID-5/6 Saving
	Minimum Host Required	Total Capacity Requirement	Minimum Host Required	Total Capacity Requirement	
0	3	1x	N/A	N/A	N/A
1	3	2x	4	1.33x	33% less
2	5	3x	6	1.5x	50% less
3	7	4x	N/A	N/A	N/A

Table 4.2 – The summary of FTMs and FTT on the vSAN storage policy

> **Important Note**
>
> You can refer to this link for details of the stripe width improvements in vSAN 7 update 1:
> `https://blogs.vmware.com/virtualblocks/2021/01/21/stripe-width-improvements-in-vsan-7-u1/`.

The last section will discuss the benefits of VMware vSAN HCI Mesh.

VMware vSAN HCI Mesh

VMware HCI Mesh was introduced with vSphere 7.0 Update 2 and allowed you to remotely connect to one or more VMware vSAN datastores from another vSAN cluster. In *Figure 4.12*, there are three VxRail clusters (one server and two clients). Two VxRail clusters (**Client**) are remotely connected to a vSAN datastore from the VxRail cluster (**Server**). vSAN HCI Mesh can be used for balancing storage resources by migrating VMs to other vSAN clusters using vSphere Storage vMotion.

Now, compute clusters can connect to the remote vSAN datastores. You can define the different storage tier vSAN storage policies (for example, RAID-1, RAID-5, deduplication, encryption, and All-Flash) and apply them to the different tier VMs in the vSAN cluster. HCI Mesh brings together multiple independent clusters for a native, cross-cluster architecture that disaggregates compute and storage resources and enables efficient utilization of storage capacity. HCI Mesh is used to share the storage resources of a VxRail cluster with other VxRail clusters or other independent clusters:

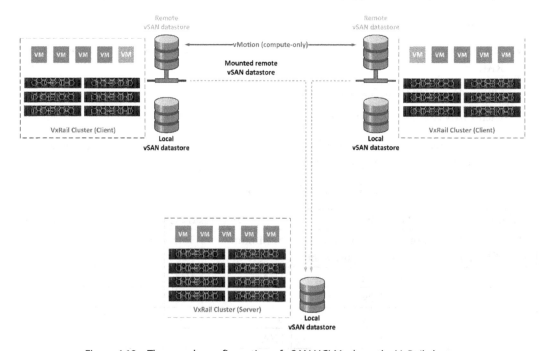

Figure 4.12 – The sample configuration of vSAN HCI Mesh on the VxRail cluster

Now, we will discuss two scenarios to learn about the benefits of VMware vSAN HCI Mesh on the VxRail cluster.

In *Figure 4.13*, the system administrator needs extra storage for deploying new VMs into the vSphere cluster in **Remote Data Center A**. But there are no storage resources available in the vSphere cluster and they cannot add other storage resources to the vSphere cluster. They can add new vSAN nodes into the existing VxRail All-Flash cluster in **Headquarter Data Center** and expand the storage capacity of the vSAN datastore, then remotely share the storage resources to the vSphere cluster in **Remote Data Center A**:

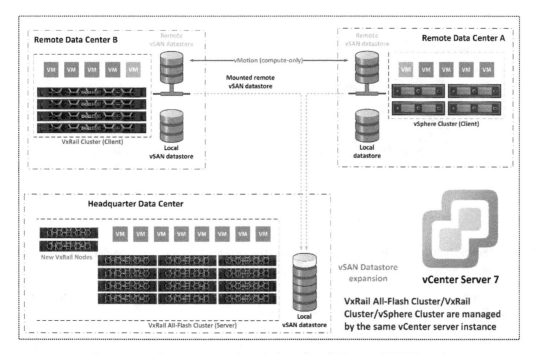

Figure 4.13 – Scenario 1 regarding the benefits of VMware vSAN HCI Mesh

In this scenario, the system administrator can easily share the storage resources in the vSphere cluster, and service interruption is not required for this configuration.

In *Figure 4.14*, **Remote Data Center A** needs to shut down due to maintenance. How do we minimize the service interruption of the VMs running in the vSphere cluster? The system administrator can migrate the VMs with VMware vMotion into the VxRail cluster in **Remote Data Center B**:

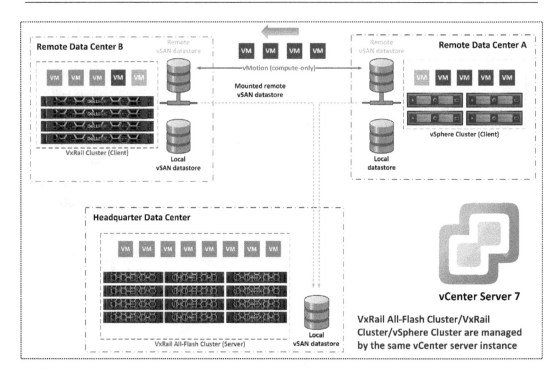

Figure 4.14 – Scenario 2 regarding the benefits of VMware vSAN HCI Mesh

In this scenario, you can easily move the VM onto the two platforms across two data centers and service interruption is not required for this configuration.

According to these two examples, you now understand the benefits of VMware vSAN HCI Mesh and how to offload the daily operation in the virtualization environment.

Please refer to *Table 4.3* for the hardware and software requirements for each scenario in *Figure 4.13* and *Figure 4.14*:

Location	Headquarter Data Center	Remote Data Center A	Remote Data Center B
Number of hosts	Eight or above	Four or above	Four or above
Number of CPU sockets and RAM on each host	Two/512 GB or above	Two/256 GB or above	Two/256 GB or above
Number of vSAN network uplinks	Two or above	Two or above	Two or above
Network speed	10 Gb or 25 Gb	10 Gb or 25 Gb	10 Gb or 25 Gb
vSphere license edition	vSphere Enterprise Plus	vSphere Standard	vSphere Standard
vSAN license edition	vSAN Enterprise or vSAN Enterprise Plus	It is not required	vSAN Enterprise or vSAN Enterprise Plus
vCenter license edition	vCenter Server Standard	vCenter Server Standard	vCenter Server Standard
Cluster type	VxRail All-Flash	vSphere Standard cluster	VxRail Hybrid
Number of vSAN disk groups	Two or above	It is not required	Two or above
vSAN Data-At-Rest Encryption	Enabled	It is not required	Disabled
Failures To Tolerate	RAID-5	It is not required	RAID-1
vSAN HCI Mesh	Enabled	N/A	N/A
Lifecycle Management	Enabled	Enabled	Enabled

Table 4.3 – The details of each scenario in Figures 4.13 and 4.14

To plan for vSAN HCI Mesh on the VxRail 7.x system, you can refer to the examples in *Figure 4.13* and *Figure 4.14*.

Summary

In this chapter, you got an overview of the design of VMware vSAN on the VxRail 7.x system, including FTMs, FTT, and SWs. You now understand the relationships of vSAN objects and components on the different vSAN storage policies – for example, RAID-1 and RAID-5. Based on the different scenarios, you understand the benefits of the vSAN storage policy and how to offload the daily operations in a virtualization environment. If the storage resource requirements on a VxRail 7.x system are ready, you can update the vSAN storage policy at any time without disrupting the service of the VMs.

In the next chapter, you will learn about VxRail cluster expansion, including VxRail scale-out rules, disk group upgrades, and node upgrades.

Questions

The following are a short list of review questions to help reinforce your learning and help you identify areas that require some improvement.

1. Which components are included in a vSAN disk group on each VxRail node?

    A. A flash drive device

    B. A 10 GB Ethernet adapter

    C. VMware vSAN licenses

    D. VMware vSphere licenses

    E. HDD devices

    F. All of the above

2. Which FTT configuration is supported on the VxRail 7.x platform?

    A. No data redundancy

    B. One failure – RAID-1 (mirroring)

    C. Two failures – RAID-1 (mirroring)

    D. One failure – RAID-5 (erasure coding)

    E. Two failures – RAID-6 (erasure coding)

    F. Three failures – RAID-1 (mirroring)

    G. All of the above

3. How many vSAN components will be created automatically if the FTM is set to RAID-1 and FTT is set to 1?

    A. 1

    B. 2

    C. 3

    D. 4

    E. 5

    F. 6

4. How many vSAN components will be created automatically if the FTM is set to RAID-1 and FTT is set to 2?

    A. 1

    B. 2

C. 3

D. 4

E. 5

F. 6

5. How many vSAN components will be created automatically if the FTM is set to RAID-5 and FTT is set to 1?

A. 1

B. 2

C. 3

D. 4

E. 5

F. 6

6. How many vSAN components will be created automatically if the FTM is set to RAID-6 and FTT is set to 2?

A. 1

B. 2

C. 3

D. 4

E. 5

F. 6

7. What is the number of minimum hosts required in the vSAN cluster if the FTM is set to RAID-1 and FTT is set to 1? This policy can be satisfied by three nodes. A fourth node is required in case of failure.

A. 3

B. 4

C. 5

D. 6

E. 7

8. What is the number of minimum hosts required in the vSAN cluster if the FTM is set to RAID-5 and FTT is set to 1? This policy can be satisfied by four nodes. The fifth node is required in case of failure.

    A. 3

    B. 4

    C. 5

    D. 6

    E. 7

9. What is the number of minimum hosts required in the vSAN cluster if the FTM is set to RAID-6 and FTT is set to 2? This policy can be satisfied by six nodes. The seventh node is required in case of failure.

    A. 3

    B. 4

    C. 5

    D. 6

    E. 7

10. Which vSAN license edition is supported if vSAN HCI Mesh is enabled on the VxRail cluster?

    A. VMware vSAN Standard edition

    B. VMware vSAN Advanced edition

    C. VMware vSAN Enterprise edition

    D. VMware vSAN Enterprise Plus edition

    E. All of the above

11. Which types of disk groups can be supported on a VxRail cluster?

    A. The All-Flash disk group only

    B. NL-SAS and SAS disk groups

    C. The Hybrid disk group only

    D. The Hybrid and All-Flash disk groups

    E. The SAS disk group only

    F. All of the above

12. Which features are available on the encryption service in the storage rules when you define the vSAN storage policy?

    A.   Data-At-Rest encryption

    B.   No encryption

    C.   No preference

    D.   Data encryption

    E.   All of the above

5

Design of Cluster Expansion

In the previous chapter, you had an overview and learned about the design of VMware vSAN on the VxRail 7.x system, including **Failure Tolerance Method (FTM)**, **Failures to Tolerate (FTT)**, and stripe width. You understood the relationship and instead of vSAN objects and components on different vSAN storage policies—for example, RAID 1 and RAID 5. Based on the different scenarios, you now understand the benefits of the vSAN storage policy of how to offload daily operations in the virtualization environment.

Scale-up and scale-out are the core features of the VxRail 7.x system. You can add different hardware components to an existing VxRail cluster if you plan to increase the resources on the VxRail 7.x system, including memory, number of capacity storage drives, and number of VxRail nodes. Each model of VxRail supports different scale-up configurations. This chapter will discuss the design of disk groups on each type of VxRail 7.x system—that is, VxRail E-Series, P-Series, V-Series, S-Series, D-Series, and G-Series.

This chapter includes the following main topics:

- Overview of VxRail scale-out rules
- Design of disk groups for various VxRail appliances – E-Series, P-Series, V-Series, S-Series, D-Series, and G-Series

Let's get started!

Overview of VxRail scale-out rules

When you deploy a VxRail cluster, the first three nodes need to be identical models—for example, a VxRail cluster with three E660 nodes. If the hardware resources (for example, memory, storage capacity, network uplinks, and so on) are not enough for the **virtual machines (VMs)** running in the VxRail cluster, you can scale up or scale out the VxRail cluster. If you scale out the VxRail cluster, you need to consider the following VxRail scale-out rules:

- All nodes in a VxRail cluster must run the same version of VxRail software—for example, VxRail Manager, VMware vSphere, hardware, firmware, Ethernet driver, or **host bus adapter** (**HBA**) driver. VxRail appliances have a lot more components, and all of them should/would be identical.

- All-Flash or **Non-Volatile Memory Express** (**NVMe**) nodes cannot be mixed in a VxRail hybrid cluster.

- All VxRail G-Series nodes must be identical in their configuration in a chassis.

- 10 GB and 1 GB networks cannot be used together for vSAN traffic; a 10 GB network is the recommended setting.

- Neither 10 GB nor 25 GB networks can be used for vSAN traffic.

- A 1 GB network can only be used for a VxRail hybrid cluster with a single processor.

- All VxRail nodes must run the same base network speed, that is, 25 GB, 10 GB, or 1 GB.

- The mixing of different VxRail series in the same cluster is supported; however, the processors must be from the same vendor.

- The mixing of All-Flash nodes and NVMe nodes in the same cluster is supported.

- VxRail nodes and VxRail dynamic nodes cannot be supported in the same cluster.

- The VxRail G-Series platform can be partially populated.

- The maximum number of VxRail nodes per cluster is 64.

- A two-node vSAN configuration is required with a witness virtual appliance; scale-out is not supported on a two-node vSAN cluster.

The next section will discuss two scenarios for VxRail scale-out.

Scenario 1

In *Figure 5.1*, there is a VxRail All-Flash cluster with three nodes (**VxRail E660F A/B/C**). Each node is installed with eight 3.8 TB-capacity SSDs and two 800 GB-cache SSDs. In this cluster, we configured two vSAN disk groups; the first one is disk slots 0 to 3 (capacity tier) and disk slot 8 (cache tier), and the other is disk slots 4 to 7 (capacity tier) and disk slot 9 (cache tier). This cluster is running VxRail software 7.0.240 and has applied the vSAN storage policy with FTM set to RAID 1 and FTT set to 1:

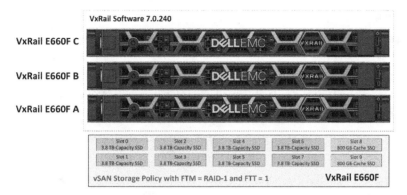

Figure 5.1 – VxRail All-Flash cluster with three nodes

We will add a new VxRail E660F node to this VxRail cluster. According to the VxRail scale-out rules, the new node must fulfill the following requirements:

- The new node is the E660F All-Flash model.

- The number of network uplinks and the speed are the same as in the existing VxRail cluster.

- The VxRail software on the new node must be the same as the existing VxRail cluster.

- The CPU model of the new node is the same as the existing VxRail nodes.

- The storage devices of the new node are the same as the existing VxRail nodes—that is, the number and type of capacity and cache devices.

In *Figure 5.2*, the total resources of this VxRail cluster will be increased automatically after adding the new node (**VxRail E660F D**), new node resources will be added to the existing cluster, and capacity will be increased. For scale-up and scale-out activities, all existing policies would be compatible with the future state of the cluster:

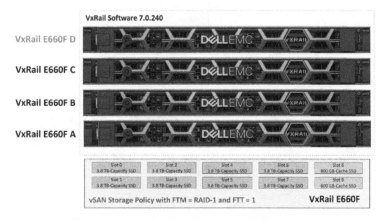

Figure 5.2 – VxRail All-Flash cluster with four nodes

> **Important Note**
> The VMware vSphere and vSAN license edition of the new VxRail node must be the same as the existing VxRail node.

This scenario is an example of a standard VxRail scale-out operation. We will now discuss another scenario of VxRail scale-out in the next section.

Scenario 2

In *Figure 5.3*, there is a VxRail All-Flash cluster with three nodes (**VxRail E660F A/B/C**). Each node installed four 3.8 TB-capacity SSDs and two 800 GB-cache SSDs. This cluster configuration has one vSAN disk group: disk slots 0 to 3 (capacity tier) and disk slot 8 (cache tier). This cluster runs in VxRail software 7.0.240 and applies the vSAN storage policy with FTM set to RAID-1 and FTT set to 1:

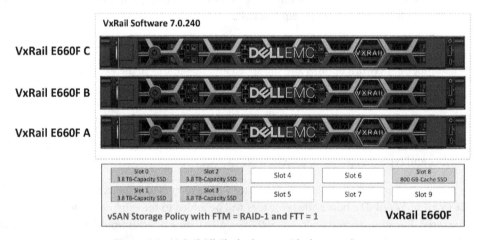

Figure 5.3 – VxRail All-Flash cluster with three nodes

In this scenario, we will add two new VxRail E660F nodes to the VxRail cluster. You can change the FTM settings at any time if the storage resources can fulfill the requirement of your new FTM settings. This operation does not interrupt the operation of the VMs in the VxRail cluster; this is one of the key features of the VxRail system.

In *Figure 5.4*, the total resources of this VxRail cluster will be increased automatically after adding the new nodes (**VxRail E660F D/E**) to the existing VxRail cluster, including CPU, memory, and storage resources. The vSAN storage policy with FTM set to RAID-1 and FTT set to 1 is still valid in this VxRail cluster. When the cluster expansion is successfully completed, then you can change the FTM to RAID-5:

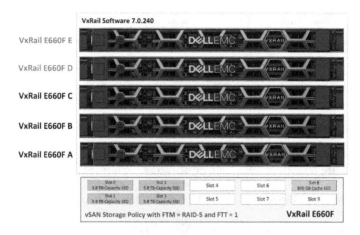

Figure 5.4 – VxRail All-Flash cluster with five nodes

You can see from the examples in *Figure 5.2* and *Figure 5.4* that the VxRail scale-out operation is very simple and flexible. The following sections will discuss the disk group design of each VxRail model.

Design of disk groups on VxRail E-Series

The cache and capacity disks are predefined in specific disk slots before the VxRail system is delivered to the customer from the Dell factory. The disk slot locations for cache and capacity are fixed when they are made in the Dell factory, and you cannot change them. *Figure 5.5* shows the front view of VxRail **E660**, **E660F**, and **E660N**. There are 10 disk slots on these 3 types of VxRail E-Series; disk slots 0 to 7 are used for the capacity tier and disk slots 8 to 9 are used for the cache tier. The capacity disks support SAS/SATA/SSD, and the cache disks support SSD and NVMe SSD:

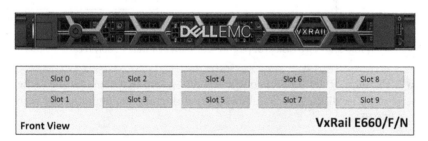

Figure 5.5 – Front view of VxRail E660/F/N

If you scale up the VxRail cluster, you need to consider the following VxRail scale-up rules:

- Mixing the SAS/SATA/NVMe SSDs in the same disk group is not supported.

- Mixing the capacity HDDs with capacity SSDs in the same disk group is not supported.

- All the capacity disks must be of the same capacity in the same disk group.

- Having different capacity disk sizes and types in the different disk groups in a VxRail node is supported.

- Having different cache disk sizes and types in the different disk groups in a VxRail node is supported.

- In VxRail software 7.0.200 or later, a large-capacity disk can be used when replacing a disk in the disk group.

- In VxRail software 7.0.200 or later, a large-capacity disk can be used when expanding a disk group.

In the release of VxRail 7.0.201 and later, VxRail E-Series supports two options of disk groups:

- **Option 1**: VxRail E-Series supports two vSAN disk groups, which contain one cache disk and up to four capacity disks per disk group. *Table 5.1* shows a two-disk-groups configuration for each disk slot in VxRail E660/F/N:

Disk Groups	Cache Tier	Capacity Tier
Disk Group 1	Slot 8	Slot 0
		Slot 1
		Slot 2
		Slot 3
Disk Group 2	Slot 9	Slot 4
		Slot 5
		Slot 6
		Slot 7

Table 5.1 – Two disk groups configuration in VxRail E660/F/N

- **Option 2**: VxRail E-Series supports one vSAN disk group that contains one cache disk and up to seven capacity disks. *Table 5.2* shows a one disk group configuration for each disk slot in VxRail E660/F/N:

Disk Groups	Cache Tier	Capacity Tier
Disk Group 1	Slot 8	Slot 0
		Slot 1
		Slot 2
		Slot 3
		Slot 4
		Slot 5
		Slot 6

Table 5.2 – One disk group configuration in VxRail E660/F/N

If you choose the one disk group configuration in VxRail E-Series, two disk slots (slot 7 and slot 9) are useless (refer to the details in *Figure 5.6*):

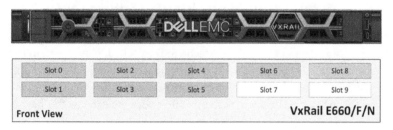

Figure 5.6 – Disk layout for one disk group configuration in VxRail E660/F/N

In VxRail E660 and E660F, you can choose SAS, SATA, or SSD for the capacity tier. For the cache tier, you can choose SSD only. VxRail E660N only supports capacity NVMe for the capacity tier and cache NVMe for the cache tier. Dell's 15th-generation PowerEdge server has new models of VxRail E-Series: E665, E665F, and E665N. The disk group configuration of E665N is the same as the E660N. The disk group configuration of E665 and E665F is different from E660 and E660F because there are only eight disk slots in VxRail E665 and E665F. Both models also support two disk groups, each disk group with one cache disk and up to three capacity disks (refer to *Figure 5.7*):

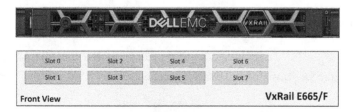

Figure 5.7 – Front view of VxRail E665/F

Table 5.3 shows the two disk groups configuration for each disk slot in VxRail E665/F:

Disk Groups	Cache Tier	Capacity Tier
Disk Group 1	Slot 6	Slot 0
		Slot 1
		Slot 2
Disk Group 2	Slot 7	Slot 3
		Slot 4
		Slot 5

Table 5.3 – Two disk groups configuration in VxRail E665/F

The next section will discuss an unsupported scale-up scenario on VxRail E-Series.

An unsupported scenario of scale-up on VxRail E-Series

In *Figure 5.8*, there is a VxRail All-Flash cluster with three nodes (**VxRail E660F A/B/C**). Each node has four 3.8 TB-capacity SSDs and one 800 GB-cache SSD installed. Now, we plan to add two 3.8 TB-capacity SSDs (**slots 4** and **5**) to expand the disk group in each VxRail E660F node, but this is not a supported scale-up configuration. Because the number of capacity disks supported is up to four, if you want to expand the disk group on each VxRail E660F node, you can create a new disk group in each node:

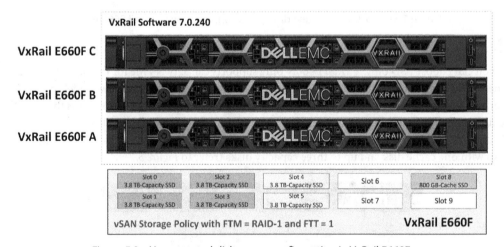

Figure 5.8 – Unsupported disk group configuration in VxRail E660F

The next section will discuss a supported scenario of scale-up on VxRail E-Series.

A supported scenario of scale-up on VxRail E-Series

Figure 5.9 shows a supported disk group configuration on VxRail E-Series. Each node has eight 3.8 TB-capacity SSDs and two 800 GB-cache SSDs installed. There are two disk groups installed; each disk group contains four 3.8 TB-capacity SSDs and one 800 GB-cache SSD. The new disk group is created by **slots 4 to 7** and **9**; this is a supported scenario:

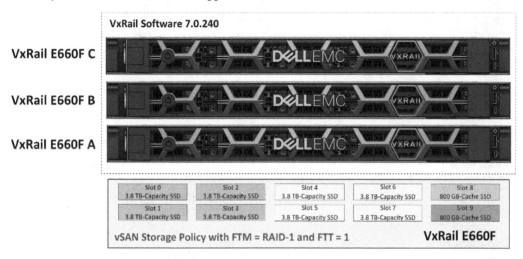

Figure 5.9 – Supported disk group configuration in VxRail E660F

According to *Table 5.3*, you learned about the disk group design for VxRail E-Series. The next section will discuss the disk group design for VxRail P-Series.

Design of disk groups on VxRail P-Series

Figure 5.10 shows the front view of VxRail P670F. There are 28 disk slots; disk slots 0 to 19 and slots 24 to 27 are used for the capacity tiers, and disk slots 20 to 23 are used for the cache tier. The capacity disks support SAS/SSD, and the cache disks support SSD and NVMe SSD:

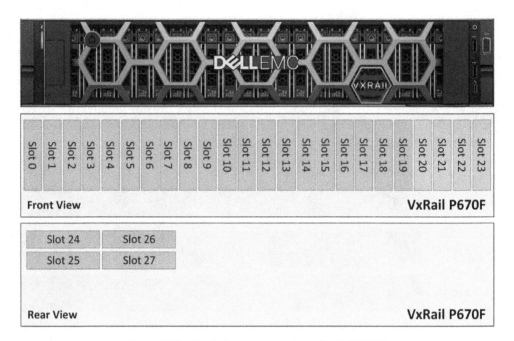

Figure 5.10 – Front view and rear view of VxRail P670F

> **Important Note**
> In Dell's 15th-generation PowerEdge server, only three models of the All-Flash node exist in the VxRail P-Series; they are VxRail P670F, P675F, and P675N.

In the release of VxRail 7.0.201 and later, VxRail P670F supports two options of disk groups:

- **Option 1**: The first configuration is of four vSAN disk groups, which contain one cache disk and up to five capacity disks per disk group. *Table 5.4* shows a four-disk-groups configuration for each disk slot in VxRail E660/F/N:

Disk Groups	Cache Tier	Capacity Tier
Disk Group 1	Slot 20	Slots 0 to 4
Disk Group 2	Slot 21	Slots 5 to 9
Disk Group 3	Slot 22	Slots 10 to 14
Disk Group 4	Slot 23	Slots 15 to 19

Table 5.4 – The first option of disk group configuration in VxRail P670F

- **Option 2**: The second configuration is of four disk groups, which contain one cache disk and up to six capacity disks per disk group. *Table 5.5* shows a four-disk-groups configuration for each disk slot in VxRail E660/F/N:

Disk Groups	Cache Tier	Capacity Tier
Disk Group 1	Slot 20	Slots 0 to 4 and Slot 24
Disk Group 2	Slot 21	Slots 5 to 9 and Slot 25
Disk Group 3	Slot 22	Slots 10 to 14 and Slot 26
Disk Group 4	Slot 23	Slots 15 to 19 and Slot 27

Table 5.5 – The second option of disk groups configuration in VxRail P670F

In Dell's 15th-generation PowerEdge server, there are two new models of VxRail P-Series: P675F and P675N. In *Figure 5.11*, the total number of disk slots of **P675F/N** is less than **P670F**, with only 24 disk slots. This model can support up to four disk groups:

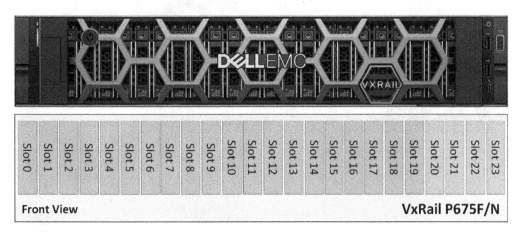

Figure 5.11 – Front view and rear view of VxRail P675F/N

Table 5.6 shows a four-disk-groups configuration for each disk slot in VxRail P675F/N:

Disk Groups	Cache Tier	Capacity Tier
Disk Group 1	Slot 20	Slots 0 to 4
Disk Group 2	Slot 21	Slots 5 to 9
Disk Group 3	Slot 22	Slots 10 to 14
Disk Group 4	Slot 23	Slots 15 to 19

Table 5.6 – Four disk groups configuration in VxRail P675F/N

The next section will discuss an unsupported scale-up scenario on VxRail P-Series.

An unsupported scenario of scale-up on VxRail P-Series

In *Figure 5.12*, there is a VxRail All-Flash cluster with three nodes (**VxRail P670F A/B/C**). There are 28 disk slots on each VxRail P670F node, configured in three disk groups. Each disk group includes one cache SSD and five capacity disks; disk slots 0 to 19 and slots 24 to 27 are used for the capacity tiers, and disk slots 20 to 23 are used for the cache tier. Now, you plan to add one capacity disk to expand each disk group; the new capacity disks are installed on **slots 15**, **16**, and **17**. But it is an unsupported disk group configuration; disk slots 15 to 19 are used for the capacity tier of the fourth disk group. You need to install the new capacity into disk slots 24 to 26:

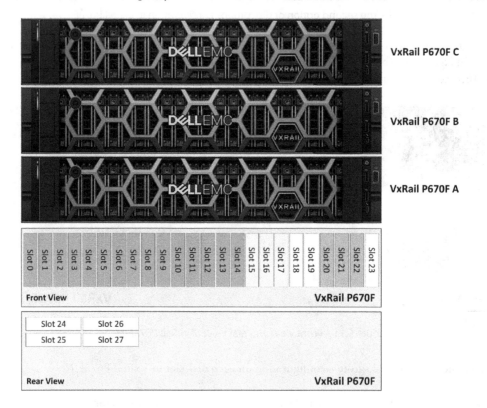

Figure 5.12 – Unsupported disk group configuration in VxRail P670F

The next section will discuss a supported scenario of scale-up on VxRail P-Series.

A supported scenario of scale-up on VxRail P-Series

Figure 5.13 shows a supported disk group configuration on VxRail P-Series. Each node has twenty capacity SSDs and four cache SSDs installed. There are four disk groups installed: each disk group contains five capacity SSDs and one cache SSD. The new disk group is created by **slots 15 to 19** and **23**; this is a supported scenario:

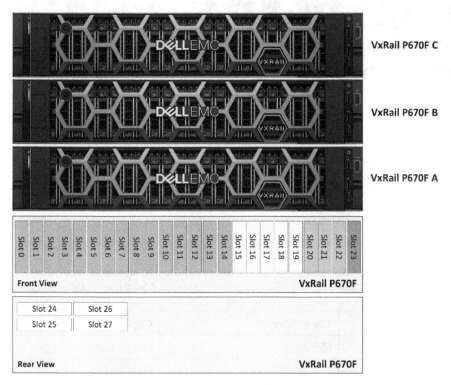

Figure 5.13 – Supported disk group configuration in VxRail P670F

According to *Tables 5.4* and *5.5*, you now know the disk group design for all VxRail P-Series systems. The next section will discuss the disk group design for VxRail V-Series.

Design of disk groups on VxRail V-Series

Figure 5.14 shows the front view of VxRail V670F. There are 24 disk slots on VxRail V670F; disk slots 0 to 19 are used for the capacity tiers, and disk slots 20 to 23 are used for the cache tier. The capacity disks support SAS/SATA/SSD, and the cache disks support SSD and NVMe SSD:

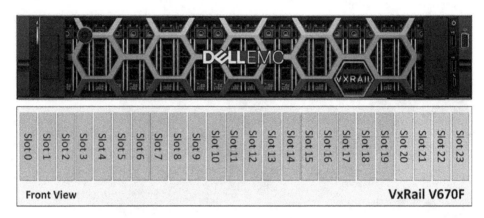

Figure 5.14 – Front view of VxRail V670F

In the release of VxRail 7.0.201 and later, VxRail V670F supports two options of disk groups:

- **Option 1**: Four vSAN disk groups, which contain one cache disk and up to five capacity disks per disk group. *Table 5.7* shows a four-disk-group configuration for each disk slot in VxRail V670F:

Disk Groups	Cache Tier	Capacity Tier
Disk Group 1	Slot 20	Slots 0 to 4
Disk Group 2	Slot 21	Slots 5 to 9
Disk Group 3	Slot 22	Slots 10 to 14
Disk Group 4	Slot 23	Slots 15 to 19

Table 5.7 – The first option of disk groups configuration in VxRail V670F

- **Option 2**: In *Figure 5.15*, the three vSAN disk groups contain one cache disk and up to five capacity disks per disk group:

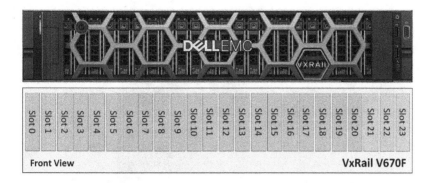

Figure 5.15 – The second option of disk groups configuration in VxRail V670F

Table 5.8 shows a three-disk-groups configuration for each disk slot in VxRail V670F:

Disk Groups	Cache Tier	Capacity Tier
Disk Group 1	Slot 21	Slots 0 to 6
Disk Group 2	Slot 22	Slots 7 to 13
Disk Group 3	Slot 23	Slots 14 to 20

Table 5.8 – Three disk groups configuration for each disk slot in VxRail V670F

The next section will discuss an unsupported scale-up scenario on VxRail V-Series.

An unsupported scenario of scale-up on VxRail V-Series

In *Figure 5.16*, there is a VxRail All-Flash cluster with three nodes (**VxRail V670F A/B/C**). There are 24 disk slots on each VxRail V670F and configured in 1 disk group, which includes 1 cache SSD and 5 capacity SSDs; disk slots 0 to 19 and slots 24 to 27 are used for the capacity tiers, and disk slots 20 to 23 are used for the cache tier. Now, you plan to add two capacity SAS disks to expand this disk group; the new capacity disks are installed on **slots 5** and **6**. But it is an unsupported disk group configuration, because the type of existing capacity disk is SDD, and the type of the new disk is SAS. The disk type of capacity disks must be the same:

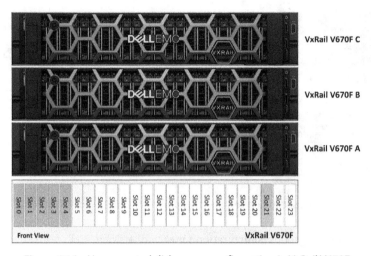

Figure 5.16 – Unsupported disk group configuration in VxRail V670F

The next section will discuss a supported scenario of scale-up on VxRail V-Series.

A supported scenario of scale-up on VxRail V-Series

Figure 5.17 shows a supported disk group configuration on VxRail V-Series. Each node has 10 capacity SSDs and 2 cache SSDs installed. There are two disk groups installed; each disk group contains five capacity SSDs and one cache SSD. The new disk group is created by **slots 7 to 11** and **22**; this is a supported scenario:

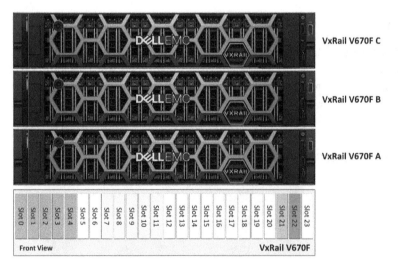

Figure 5.17 – Supported disk group configuration in VxRail V670F

According to *Tables 5.7* and *5.8*, you now know the disk group design for VxRail V-Series. The next section will discuss the disk group design for VxRail S-Series.

Design of disk groups on VxRail S-Series

This section will discuss the disk group design for VxRail V-Series. *Figure 5.18* shows the front view of VxRail S670. There are 16 disk slots on VxRail S670; disk slots 0 to 11 (front view) are used for the capacity tiers, and disk slots 0 to 3 (rear view) are used for the cache tier. VxRail S-Series only supports a hybrid configuration. The capacity disks only support 3.5" SAS/SATA, and the cache disks only support 2.5" SSD or NVMe:

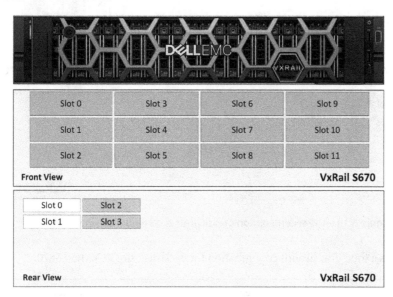

Figure 5.18 – The first option of disk groups configuration in VxRail S670

In the release of VxRail 7.0.201 and later, VxRail S670 supports two options of disk groups:

- **Option 1**: The two vSAN disk groups, which contain one cache disk and up to six capacity disks per disk group. *Table 5.9* shows a four-disk-groups configuration for each disk slot in VxRail S670:

Disk Groups	Cache Tier	Capacity Tier
Disk Group 1	Slot 2 (rear view)	Slots 0 to 5 (front view)
Disk Group 2	Slot 3 (rear view)	Slots 6 to 11 (front view)

Table 5.9 – The first option of disk groups configuration in VxRail S670

- **Option 2**: *Figure 5.19* shows the four vSAN disk groups, which contain one cache disk and up to three capacity disks per disk group:

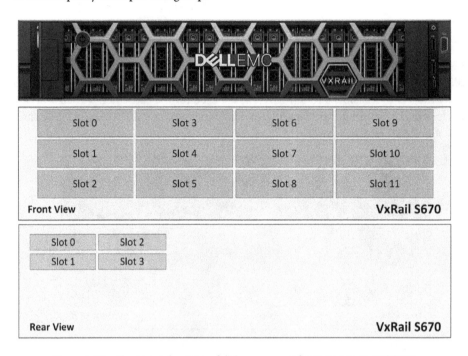

Figure 5.19 – The second option of disk groups configuration in VxRail S670

Table 5.10 shows a four-disk-groups configuration for each disk slot in VxRail S670:

Disk Groups	Cache Tier	Capacity Tier
Disk Group 1	Slot 0 (rear view)	Slot 0 (front view)
		Slot 1 (front view)
		Slot 2 (front view)
Disk Group 2	Slot 1 (rear view)	Slot 3 (front view)
		Slot 4 (front view)
		Slot 5 (front view)
Disk Group 3	Slot 2 (rear view)	Slot 6 (front view)
		Slot 7 (front view)
		Slot 8 (front view)

Disk Group 4	Slot 3 (rear view)	Slot 9 (front view)
		Slot 10 (front view)
		Slot 11 (front view)

Table 5.10 – The second option of disk groups configuration in VxRail S670

The next section will discuss an unsupported scale-up scenario on VxRail S-Series.

An unsupported scenario of scale-up on VxRail S-Series

In *Figure 5.20*, there is a VxRail hybrid cluster with three nodes (**VxRail S670 A/B/C**). There are 16 disk slots on each VxRail S670 and configured in one disk group, which includes one cache SSD (rear view, **slot 3**) and three capacity HDDs (front view, slots 0 to 2); **disk slots 0 to 11** (front view) are used for the capacity tiers, and disk slots 0 to 3 (rear view) are used for the cache tier. Now, you plan to create a new disk group on each node, which includes six capacity HDDs (front view, **slots 3 to 8**) and one cache SSD (rear view, **slot 2**). But it is an unsupported disk group configuration because the new disk group only supports up to three capacity HDDs:

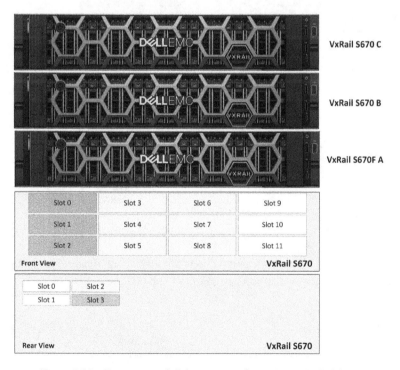

Figure 5.20 – Unsupported disk group configuration in VxRail S670

The next section will discuss a supported scenario of scale-up on VxRail S-Series.

A supported scenario of scale-up on VxRail S-Series

Figure 5.21 shows a supported disk group configuration on VxRail S-Series. Each node has six capacity disks and two cache SSDs installed. There are two disk groups installed; each disk group contains three capacity SSDs and one cache SSD. The new disk group is created by **slots 3 to 5** (front view) and **slot 1** (rear view); this is a supported scenario:

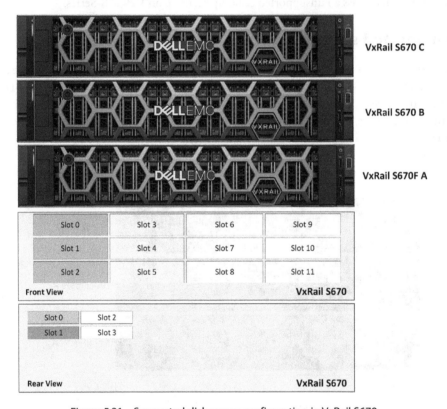

Figure 5.21 – Supported disk group configuration in VxRail S670

According to *Tables 5.9* and *5.10*, you learned about the disk group design for VxRail S-Series. The next section will discuss the disk group design for VxRail D-Series.

Design of disk groups on VxRail D-Series

Figure 5.22 shows the front view of VxRail D560/F. There are only eight disk slots on VxRail D560/F; disk slots 0 to 5 are used for the capacity tiers, and disk slots 6 to 7 are used for the cache tier. The capacity disks support SAS/SSD, and the cache disks support SSD:

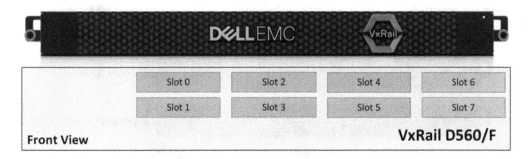

Figure 5.22 – Front view of VxRail D560/F

In the release of VxRail 7.0.201 or later, VxRail D560/F only supports one option of a disk group, which contains one cache disk and up to three capacity disks per disk group. *Table 5.11* shows a disk group configuration for each disk slot in VxRail D560/F:

Disk Groups	Cache Tier	Capacity Tier
Disk Group 1	Slot 6	Slot 0
		Slot 1
		Slot 2
Disk Group 2	Slot 7	Slot 3
		Slot 4
		Slot 5

Table 5.11 – Two disk groups configuration in VxRail D560/F

The next section will discuss an unsupported scale-up scenario on VxRail D-Series.

An unsupported scenario of scale-up on VxRail D-Series

In *Figure 5.23*, there is a VxRail hybrid cluster with three nodes (**VxRail D560 A/B/C**). There are eight disk slots on each VxRail D560F node and configured in one disk group, which includes one cache SSD (**slot 7**) and three capacity SSDs (**slots 0 to 2**); disk slots 0 to 5 are used for the capacity tiers, and disk slots 6 to 7 are used for the cache tier. Now, you plan to add one cache SSD to slot 6, but it is an unsupported disk group configuration because slot 6 is used for cache SSD for the second disk group on D560F:

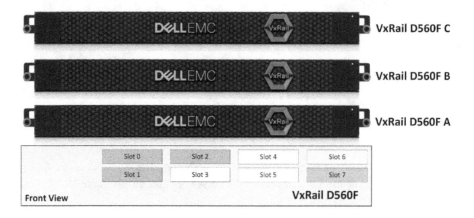

Figure 5.23 – Unsupported disk group configuration in VxRail D560F

The next section will discuss a supported scenario of scale-up on VxRail D-Series.

A supported scenario of scale-up on VxRail D-Series

Figure 5.24 shows a supported disk group configuration on VxRail D-Series. Each node has six capacity disks and two cache SSDs installed. There are two disk groups installed; each disk group contains three capacity SSDs and one cache SSD. The new disk group is created by **slots 3 to 5** and **slot 7**; this is a supported scenario:

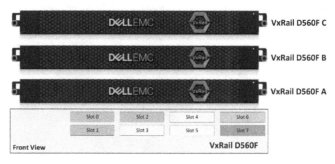

Figure 5.24 – Supported disk group configuration in VxRail D560F

According to *Table 5.11*, you know the disk group design for VxRail D-Series. The next section will discuss the disk group design for VxRail G-Series.

Design of disk groups on VxRail G-Series

This last section will discuss the disk group design for VxRail G-Series. *Figure 5.25* shows the front view and rear view of VxRail G560/F. There are 24 disk slots on the VxRail G-Series chassis; disk slots 1 to 5, 7 to 11, 13 to 17, and 19 to 23 are used for the capacity tiers, and disk slots 0, 6, 12, and 18 are used for the cache tier. It supports the installation of four nodes into each VxRail G-Series chassis. VxRail G-Series supports hybrid and All-Flash configurations. The capacity disks only support 3.5" SAS/SATA, and the cache disks only support 2.5" SSD or NVMe:

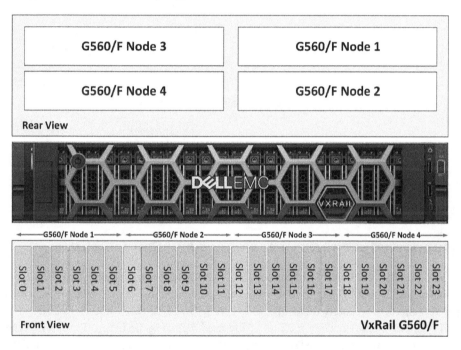

Figure 5.25 – Front view and rear view of VxRail G560/F

In the release of VxRail 7.0.201 or later, VxRail G560/F only supports one disk group.

Each disk group contains one cache disk and up to five capacity disks per disk group. *Table 5.12* shows a disk group configuration for each disk slot in the VxRail G-Series chassis:

	Node ID	Disk Groups	Cache Tier	Capacity Tier
G-Series Chassis	Node 1	Disk Group 1	Slot 0	Slots 1 to 5
	Node 2	Disk Group 1	Slot 6	Slots 7 to 11
	Node 3	Disk Group 1	Slot 12	Slots 13 to 17
	Node 4	Disk Group 1	Slot 18	Slots 19 to 23

Table 5.12 – Disk groups configuration in VxRail G560/F

Important Note

All VxRail G-Series nodes in a chassis must be identical configurations.

The next section will discuss an unsupported scale-up scenario on VxRail G-Series.

An unsupported scenario of scale-up on VxRail G-Series

In *Figure 5.26*, there is a VxRail hybrid cluster with four nodes (**VxRail G560 A/B/C/D**). There are six disk slots on each VxRail D560 node and these are configured in one disk group, which includes one cache SSD and three capacity disks; disk slots 1 to 5, 7 to 11, 13 to 17, and 19 to 23 are used for the capacity tiers, and disk slots 0, 6, 12, and 18 are used for the cache tier in the G-Series chassis. Now, you plan to add two capacity SSDs to each G560 node (slots 4, 5, 10, 11, 16, 17, 22, and 23, but it is an unsupported disk group configuration. Because this VxRail cluster is a hybrid configuration, it is not supported to change to an All-Flash configuration:

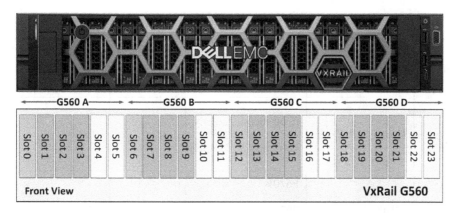

Figure 5.26 – Unsupported disk group configuration in VxRail G560

The next section will discuss a supported scenario of scale-up on VxRail G-Series.

A supported scenario of scale-up on VxRail G-Series

Figure 5.27 shows a supported disk group configuration on VxRail G-Series. Each node has six capacity disks and one cache SSDs installed. There is one disk group installed. The new capacity disks are created in each node in slots 5, 11, 17, and 23; this is a supported scenario:

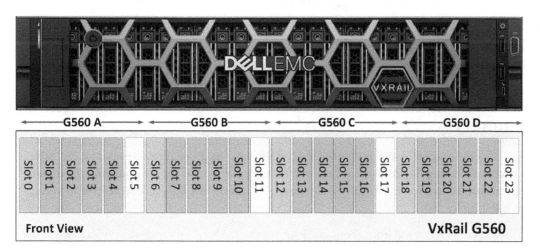

Figure 5.27 – Supported disk group configuration in VxRail G560

According to *Table 5.12*, you know the disk group design for VxRail G-Series.

Lastly, *Table 5.13* shows a summary of disk group configurations for each VxRail series:

VxRail Model	Cluster Type	Model	Form Factor	Number of Supported Disk Groups per Node
E-Series	Hybrid, All-Flash	E660, E660F, E660N E665, E665F, E665N	1 unit	2
P-Series	All-Flash	P670F P675F, P675N P580N	2 units	4
V-Series	All-Flash	V670F	2 units	4
S-Series	Hybrid	S670	2 units	4
D-Series	Hybrid, All-Flash	D560, D560F	1 unit	2
G-Series	Hybrid, All-Flash	G560, G560F	2 units per chassis	1

Table 5.13 – Summary of disk group configurations for each VxRail series

Summary

In this chapter, you covered an overview and learned about the design of cluster expansion in the VxRail 7.x system, including the scale-out and scale-in rules for each model for a VxRail node. Compared to the traditional architecture of server and storage, you can see that VxRail cluster expansion is very easy and flexible.

In the next chapter, you will learn about the design of the vSAN two-node cluster on VxRail and the best practices of this solution.

Questions

The following are a short list of review questions to help reinforce your learning and help you identify areas which require some improvement.

1. Which two types of rules can be supported on the VxRail 7.x system?

    A. VxRail upgrade rules

    B. VxRail update rules

    C. VxRail scale-up rules

    D. VxRail expansion rules

    E. VxRail scale-out rules

    F. None of these

2. Which of the following is the wrong description of VxRail scale-out rules?

    A. All-Flash and NVMe nodes cannot be mixed in a VxRail Hybrid cluster.

    B. The mixing of All-Flash nodes and NVMe nodes in the same cluster is not supported.

    C. Neither 10 GB nor 25 GB networks can be used for vSAN traffic.

    D. Only a 1 GB network can be used for a VxRail hybrid cluster with a single processor.

    E. The maximum number of VxRail nodes per cluster is 64.

3. What is the maximum number of disk groups that can be supported on the VxRail E660F?

    A. One disk group

    B. Two disk groups

    C. Three disk groups

    D. Four disk groups

    E. Five disk groups

    F. Six disk groups

4. What is the maximum number of disk groups that can be supported on the VxRail P670F?

    A. One disk group

    B. Two disk groups

    C. Three disk groups

    D. Four disk groups

    E. Five disk groups

    F. Six disk groups

5. What is the maximum number of disk groups that can be supported on the VxRail V670F?

    A. One disk group

    B. Two disk groups

    C. Three disk groups

    D. Four disk groups

    E. Five disk groups

    F. Six disk groups

6. What is the maximum number of disk groups that can be supported on the VxRail S670?

    A. One disk group

    B. Two disk groups

    C. Three disk groups

    D. Four disk groups

    E. Five disk groups

    F. Six disk groups

7. What is the maximum number of disk groups that can be supported on the VxRail D560F?

    A. One disk group

    B. Two disk groups

    C. Three disk groups

    D. Four disk groups

    E. Five disk groups

    F. Six disk groups

8. What is the maximum number of disk groups that can be supported on the VxRail G560F?

    A. One disk group

    B. Two disk groups

    C. Three disk groups

    D. Four disk groups

    E. Five disk groups

    F. Six disk groups

9. Which VxRail model of the capacity disk can be supported for up to seven disks?

    A. VxRail E660

    B. VxRail V670F

    C. VxRail P670F

    D. VxRail S670

    E. VxRail G560F

    F. None of these

10. Which VxRail model only supports a hybrid configuration?

    A. VxRail E-Series

    B. VxRail P-Series

    C. VxRail V-Series

    D. VxRail S-Series

    E. VxRail D-Series

    F. VxRail G-Series

11. Which VxRail models support both hybrid and All-Flash configurations?

    A. VxRail E-Series

    B. VxRail P-Series

    C. VxRail V-Series

    D. VxRail S-Series

    E. VxRail D-Series

    F. VxRail G-Series

12. In *Figure 5.28*, the disk slot configuration belongs to which VxRail model?

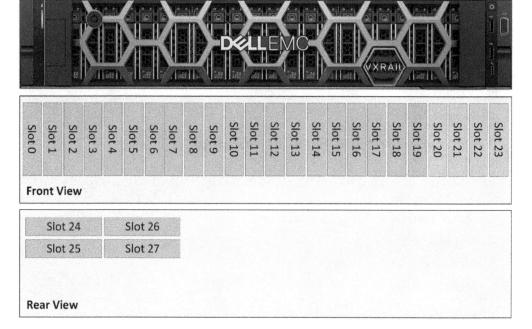

Figure 5.28 – Disk slot configuration of the VxRail 7.x system

A. VxRail E-Series

B. VxRail P-Series

C. VxRail V-Series

D. VxRail S-Series

E. VxRail D-Series

F. VxRail G-Series

Design of vSAN 2-Node Cluster on VxRail

In the previous chapter, we learned about cluster expansion in the VxRail 7.x system, including the scale-out and scale-in rules for each model for a VxRail node. Compared to the traditional server and storage architecture, VxRail cluster expansion is very easy and flexible.

Three VxRail nodes with the same hardware configuration are required in a VxRail cluster when you build a VxRail cluster prior to VxRail v4.7.100. Starting with VxRail v4.7.100, there is support for a vSAN two-node cluster on VxRail. The VMware vSAN two-node cluster on VxRail is designed for small-scale deployments, and the workloads and **High Availability** (**HA**) requirements are low. In this chapter, you will learn about VxRail vSAN two-node clusters, and we will discuss how to plan and design a VxRail vSAN two-node cluster.

This chapter includes the following main topics:

- Overview of VxRail vSAN two-node clusters
- Designing a VxRail vSAN two-node cluster
- A scenario using a VxRail vSAN two-node cluster
- Failure scenarios of a VxRail vSAN two-node cluster

Let's get started!

Overview of VxRail vSAN two-node clusters

The VMware vSAN two-node cluster on VxRail is designed for remote offices, branch offices, and small-scale deployments. The VxRail vSAN two-node configuration is supported in VxRail E, P, V, D, and S Series. This solution includes two VxRail nodes and a witness virtual appliance, and it supports the two deployment options, that is, switch configuration and direct connection configuration. In *Figure 6.1*, there are two nodes (**VxRail Node 1** and **VxRail Node 2**). There are four 10 GB ports on each node. The first and second ports (**P1** and **P2**) are used for the management, witness, and virtual machine networks. The third and fourth ports (**P3** and **P4**) are used for vSAN and vMotion networks. **vCenter Server** and **vSAN Witness** are hosted in the main data center:

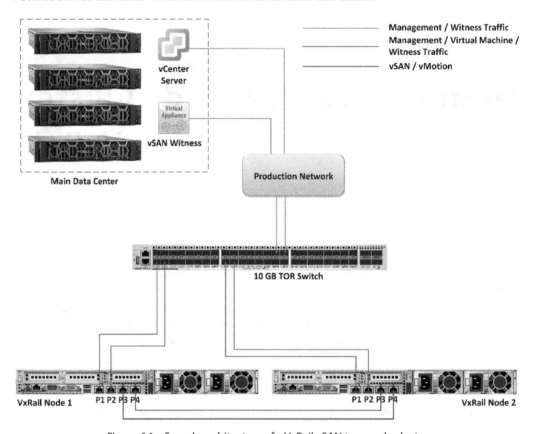

Figure 6.1 – Sample architecture of a VxRail vSAN two-node cluster

The vSAN two-node cluster is created with two VxRail nodes and the **vSAN Witness** virtual appliance. The management of this cluster is performed by vCenter Server in the main data center.

If you choose to deploy a VxRail vSAN two-node cluster, you need to consider the following:

- In VxRail v4.7.100 or later, vSAN two-node clusters support the deployment of a direct connection.

- From VxRail v4.7.410 onward, vSAN two-node clusters support the deployment of a switch configuration.

- vSAN two-node cluster deployment is only supported by VxRail E560/F/N, E665/F/N, P570/F, P675F, V570/F, D560/F, and S570.

- Four network ports per node are required. The **Network Daughter Card** (**NDC**) supports the following configurations:

  - Two 1 Gb ports and two 10 Gb ports

  - Four 10 Gb ports

  - Four 25 Gb ports (two 25 Gb ports on the NDC and two 25 Gb ports on the **Peripheral Component Interconnect Express** (**PCIe**) adapter)

- The vSAN witness appliance can be installed at the same site as the data nodes, but it cannot be installed on the vSAN two-node cluster.

- Embedded VMware vCenter Server is not supported with a vSAN two-node cluster deployment. It only supports external vCenter Server.

- The VxRail nodes must connect to vSAN Witness over a separate network. **Witness Traffic Separation** (**WTS**) is automatically configured during the vSAN two-node cluster deployment.

- VxRail vSAN two-node clusters support up to 25 virtual machines because the vSAN two-node cluster does not exceed 750 vSAN components. This is true for tiny witness appliance deployment, and they are good candidates for a two-node VxRail cluster. In the latest versions of vSAN, if we deploy a witness appliance with a normal or large size, we can have more VMs and vSAN components.

- A minimum of 25% of spare storage capacity is required for a VxRail vSAN two-node cluster. The FTM of the vSAN storage policy only supports RAID-1.

- The maximum supported **Round-Trip Time** (**RTT**) between a VxRail vSAN two-node cluster and vSAN Witness is 500 milliseconds.

- Two-node VxRail cluster expansion before VxRail version 7.0.130 is not possible.

We will discuss the architecture of a VxRail vSAN two-node cluster in the next section.

The architecture of a VxRail vSAN two-node cluster

A VxRail vSAN two-node cluster contains three nodes, that is, two data nodes and one witness node. In *Figure 6.2*, each node is configured as a **Fault Domain** (**FD**): **VxRail Node 1** is configured as **Fault Domain 1** (**Preferred**), and **VxRail Node 2** is configured as **Fault Domain 2** (**Secondary**). vSAN Witness is configured as **Fault Domain 3**. When three FDs are available, the virtual machines created on the VxRail vSAN two-node cluster have mirror protection; that is, there is one copy of data on VxRail Node 1 and a second copy on VxRail Node 2. The witness component is placed on the vSAN Witness:

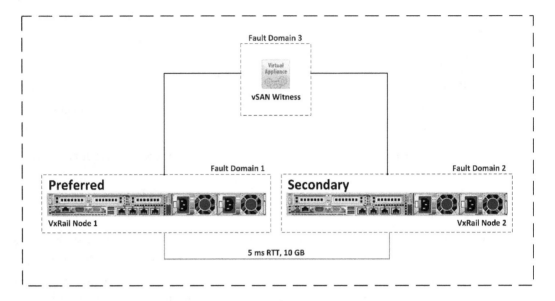

Figure 6.2 – Three FDs of a VxRail vSAN two-node cluster

> **Important Note**
>
> Starting with VxRail 4.7.300, the **Life Cycle Management** (**LCM**) of VxRail also supports the upgrading of vSAN witness components.

A vSAN license can be used on a VxRail vSAN two-node cluster, but some features are not supported. *Table 6.1* shows a feature comparison of each vSAN license edition:

vSAN Edition Features	Standard	Advanced	Enterprise	Enterprise Plus
Storage policy-based management	Supported	Supported	Supported	Supported
vSphere Distributed Switch (vDS)	Supported	Supported	Supported	Supported
FD	Supported	Supported	Supported	Supported
Software checksum	Supported	Supported	Supported	Supported
All-Flash hardware	Supported	Supported	Supported	Supported
iSCSI target service	Supported	Supported	Supported	Supported
QoS	Supported	Supported	Supported	Supported
Cloud-native storage	Supported	Supported	Supported	Supported

Deduplication and compression	N/A	Supported	Supported	Supported
RAID-5/6 erasure coding	N/A	Supported	Supported	Supported
vRealize Operations with vCenter	N/A	Supported	Supported	Supported
Data-at-rest encryption	N/A	N/A	Supported	Supported
Stretched cluster with local failure protection	N/A	N/A	Supported	Supported

Table 6.1 – A feature comparison of each vSAN license edition

Important Note

The VxRail vSAN two-node cluster expansion is not supported before VxRail version 7.0.130. Cluster expansion is not supported if the two nodes are direct connections.

Table 6.2 shows the network traffic settings of a VxRail vSAN two-node cluster. In this scenario, P1 to P4 are from the same network switch:

Network Traffic	Network I/O Control Shares	P1 on VxRail Node	P2 on VxRail Node	P3 on VxRail Node	P4 on VxRail Node	Remark
Management network	40%	Standby	Active	Unused	Unused	The VLAN is the same as the vCenter Server management network.
vCenter Server management network	N/A	Standby	Active	Unused	Unused	The VLAN is the same as the management network.
VxRail management network	N/A	Standby	Active	Unused	Unused	The VLAN ID is 3939.
vSAN network	100%	Unused	Unused	Active	Standby	N/A
vMotion network	50%	Unused	Unused	Standby	Active	N/A
Witness traffic	N/A	Active	Standby	Unused	Unused	N/A
Virtual machines	60%	Active	Standby	Unused	Unused	N/A

Table 6.2 – The network traffic settings of a VxRail vSAN two-node cluster

In the next section, we will discuss the deployment options of a VxRail vSAN two-node cluster on VxRail.

Central management and localized witness

VxRail vSAN two-node clusters support three types of deployment, that is, *central management and localized witness*, *localized management and witness*, and *central management and witness*. This section will cover an overview and discuss the benefits of the first deployment option.

In *Figure 6.3*, the vSAN two-node cluster is deployed at the remote site; there are three hosts (VxRail Node 1, VxRail Node 2, and **Management Host**). The vSAN witness appliance is hosted in the management host. This vSAN two-node cluster is managed with vCenter Server, which is located in the main data center:

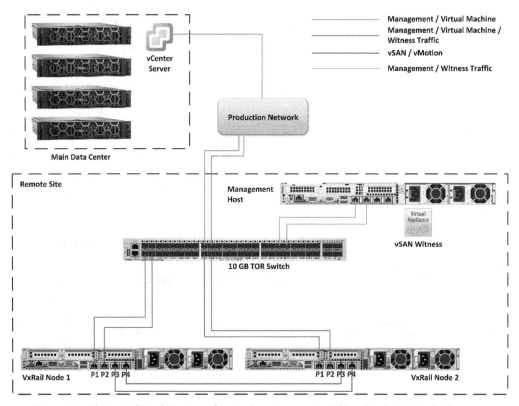

Figure 6.3 – The sample architecture of central management and localized witness deployment

Since the vSAN Witness appliance is located on the management host, the management host must access the management network, the vCenter Server management network, and witness traffic. The next section will discuss the second deployment option of a VxRail vSAN two-node cluster, localized management and witness.

Localized management and witness

In *Figure 6.4*, the vSAN two-node cluster is deployed at the remote site, and there are three hosts (VxRail Node 1, VxRail Node 2, and Management Host). The vSAN Witness appliance and vCenter Server are hosted in the management host. This VxRail vSAN two-node cluster is managed with vCenter Server at the remote site. In this type of VxRail vSAN two-node deployment, all the components are allocated in the same location:

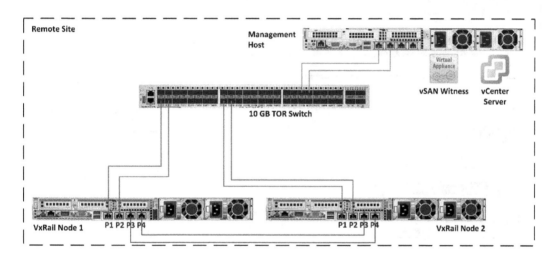

Figure 6.4 – The sample architecture of localized management and witness deployment

> **Important Note**
> vCenter Server cannot be hosted in a VxRail vSAN two-node cluster.

The following section will discuss the third deployment option of a VxRail vSAN two-node cluster, central management and witness.

Central management and witness

In *Figure 6.5*, a VxRail vSAN two-node cluster is deployed in **Remote Site A** and **Remote Site B**, and there are two hosts (VxRail Node 1 and VxRail Node 2). The vSAN Witness appliances and vCenter Server are hosted in the main data center. In this VxRail vSAN two-node deployment, one vCenter Server instance can support managing multiple VxRail vSAN two-node clusters. Each VxRail vSAN two-node cluster requires its own vSAN witness:

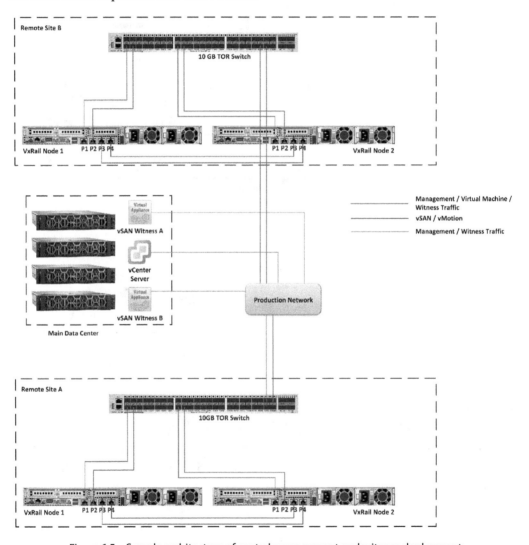

Figure 6.5 – Sample architecture of central management and witness deployment

You have now seen an overview of each VxRail vSAN two-node cluster deployment type. *Table 6.3* shows the pros and cons of each type of deployment:

Deployment Option	Pros	Cons
Central management and localized witness	It can provide a single management dashboard to manage all VxRail vSAN two-node clusters.	Extra network configuration and expenditure are required for the communication of vCenter Server and the vSAN witness. Extra software and hardware costs are required for the deployment of the vSAN witness.
Localized management and witness	Extra network configuration and expenditure are not required for the communication of vCenter Server and the vSAN witness.	Extra software and hardware expenditure are required for the deployment of multiple vCenter Server vSAN Witnesses, for example, an optional vCenter Server license per site.
Central management and witness	It can provide a single management dashboard to manage all VxRail vSAN two-node clusters. Optional vCenter Server licenses and hardware are not required.	Extra network configuration and expenditure are required for the communication of vCenter Server and vSAN Witness.

Table 6.3 – A comparison table of each deployment option of a VxRail vSAN two-node cluster

The next section will discuss planning and designing a VxRail vSAN two-node cluster.

Designing a VxRail vSAN two-node cluster

This section will discuss designing and planning a VxRail vSAN two-node cluster, including the networking and deployment type. VxRail vSAN two-node clusters support direct-connect configuration and switch configuration deployment. Both configurations require four ports (10 Gb and 25 Gb) on each VxRail node.

Direct-connect configuration

In *Figure 6.6*, you can see four 10 Gb ports in a direct-connect configuration. There are four 10 Gb ports on each VxRail node, and the **Top of Rack** (**TOR**) switch is a 10 Gb network switch. Ports 1 and 2 (**P1** and **P2**) of each VxRail node are connected to the network switch, and both ports are used for the management and vSAN Witness networks. Ports 3 and 4 (**P3** and **P4**) of VxRail Node 1 are directly connected to ports P3 and P4 of VxRail Node 2, and both ports are used for vSAN and vMotion networks:

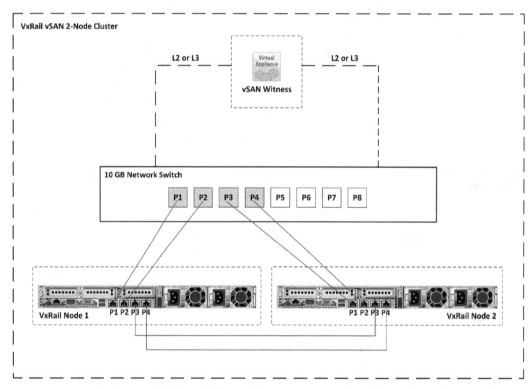

Figure 6.6 – Four 10 Gb ports in a direct-connect configuration

Now we'll discuss another example of a direct-connect configuration. If the TOR switch is a 25 Gb network switch, the network port of each VxRail node must be 25 Gb.

In *Figure 6.7*, you can see four 25 Gb ports in a direct-connect configuration. There are four 25 Gb ports on each VxRail node, and the TOR switch is a 25 Gb network switch. Ports 1 and 3 (P1 and P3) of each VxRail node are connected to the network switch, and both ports are used for the management and vSAN witness networks. Ports 2 and 4 (P2 and P4) of VxRail Node 1 are directly connected to ports 2 and 4 (P2 and P4) of VxRail Node 2, and both ports are used for vSAN and vMotion networks. Two network adapters are installed on each VxRail node; one is two 25 Gb ports NDC, and the other is two 25 Gb port PCIe network adapters. This configuration can provide the hardware with the HA network adapter on each VxRail node. The VxRail network services are not interrupted if one of the hardware adapters is faulty:

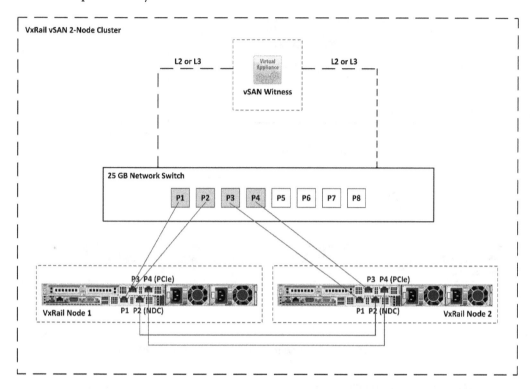

Figure 6.7 – Four 25 Gb ports in a direct-connect configuration

With the previous examples in *Figures 6.6* and *6.7*, you've learned about the architecture of a direct-connect configuration. The next section will discuss switch configuration.

Switch configuration

Now, we will discuss some examples of switch configuration on a VxRail vSAN two-node cluster. In *Figure 6.8*, you can see four 10 Gb ports in a switch configuration with a single switch. There are four 10 Gb ports on each VxRail node, and the TOR switch is a 10 Gb network switch. Ports 1 and 2 (P1 and P2) of each VxRail node are connected to the network switch (P1 to P4), and both ports are used for the management and vSAN witness networks. Ports 3 and 4 (P3 and P4) of each VxRail node are connected to the network switch (**P5** to **P8**); both ports are used for vSAN and vMotion networks. Cluster expansion is supported in this configuration:

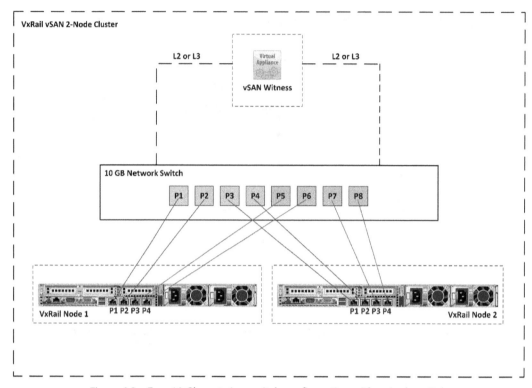

Figure 6.8 – Four 10 Gb ports in a switch configuration with a single switch

Now, we will discuss another example of a switch configuration. In *Figure 6.9*, you can see four 10 Gb ports in a switch configuration with a dual switch; there are four 10 Gb ports on each VxRail node and two 10 Gb TOR switches, which are **10 GB Network Switch A** and **10 GB Network Switch B**. Ports 1 and 2 (P1 and P2) of each VxRail node are connected to 10 GB Network Switch A and B (P1 and P3), and both ports are used for the management and vSAN witness networks. Ports 3 and 4 (P3 and P4) of each VxRail node are connected to 10 GB Network Switch A and B (P2 and P4); both ports are used for vSAN and vMotion networks. In this configuration, network switches A and B have no inter-link connections:

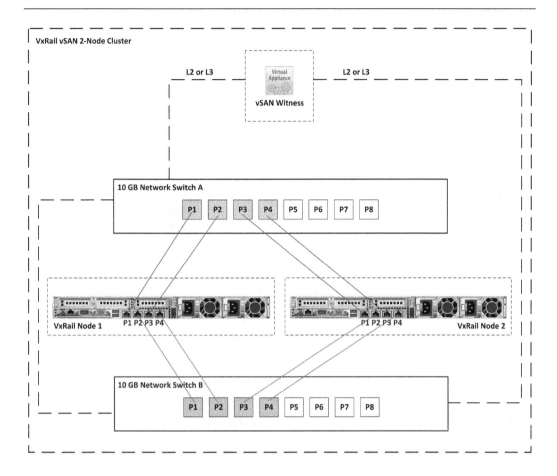

Figure 6.9 – Four 10 Gb ports in a switch configuration with dual switch

With the previous examples in *Figures 6.8* and *6.9*, you have learned about the architecture of a switch configuration.

Table 6.4 shows a comparison of direct-connect and switch configurations:

Network Deployment Option	Pros	Cons
Direct-connect configuration with an NDC	Minimizes the number of network ports and network switches for the VxRail vSAN two-node cluster configuration. No network latency for vSAN and vMotion networks across VxRail nodes.	Cluster expansion is not supported.
Direct-connect configuration with an NDC and PCIe network adapter	Minimizes the number of network ports and network switches for the VxRail vSAN two-node cluster configuration. It can provide the HA pair of network uplinks on each VxRail node.	Cluster expansion is not supported.
Single-switch configuration	Cluster expansion is supported. Single management dashboard to monitor all network traffic.	Increases the number of network ports and network switches for the VxRail vSAN two-node cluster configuration.
Dual-switch configuration	Cluster expansion is supported. Separate network switch managing the VxRail network service and vSAN network.	Increases the number of network ports and network switches for the VxRail vSAN two-node cluster configuration.

Table 6.4 – A comparison of direct-connect and switch configuration

You now understand the architecture and benefits of the direct-connect and switch configurations of a VxRail vSAN two-node cluster. When you plan to deploy the VxRail vSAN two-node cluster, you should choose the deployment option first (refer to *Table 6.3*), choosing whichever option is best for your environment. Then, you can choose either the direct-connect configuration or switch configuration, depending on what is suitable for your network requirements for deploying the VxRail vSAN two-node cluster.

The following section will discuss a sample configuration of a VxRail vSAN two-node cluster.

A scenario using a VxRail vSAN two-node cluster

This section will discuss a sample configuration of a VxRail vSAN two-node cluster, as shown in *Figure 6.10*. We will discuss a scenario of central management and witness on the VxRail vSAN two-node cluster, including network configuration and software and hardware requirements. In *Figure 6.10*, there are two separate locations, the **Main Data Center** and the **Remote Office**. vCenter Server and vSAN Witness are installed in the main data center, and two VxRail E660 appliances (VxRail Node 1 and VxRail Node 2) are installed in the remote office:

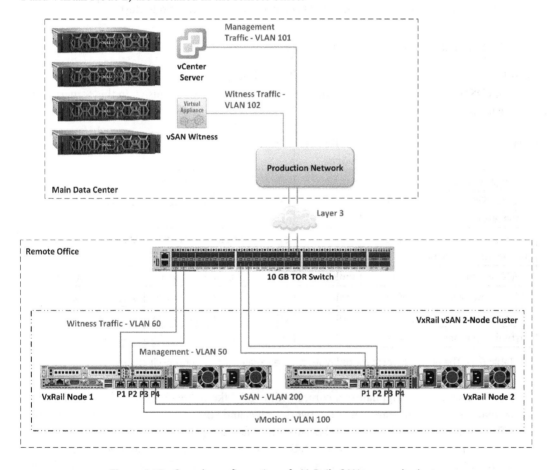

Figure 6.10 – Sample configuration of a VxRail vSAN two-node cluster

If you choose this deployment of the VxRail vSAN two-node cluster in this scenario, it must fulfill the following requirements:

- There are four 10 Gb ports on each VxRail E660.
- The vSAN witness and vCenter Server are installed in the main data center.
- The VLAN ID of vCenter Server and the vSAN witness must be different.

I've only listed the main differentiating points here. You can find the other requirements in the *Overview of VxRail vSAN two-node clusters* section of this chapter.

The next section will discuss the settings of vDS on a VxRail vSAN two-node cluster.

vSphere Distributed Switch

According to the configuration in *Figure 6.10*, VxRail will automatically create the following network groups in vDS (in *Table 6.5*) while deploying the VxRail vSAN two-node cluster. In this scenario, P1 to P4 are from the same network switch.

Network Traffic	NIOC Shares	VMkernel ports	VLAN	P1	P2	P3	P4
Management network	40%	vmk2	50	Standby	Active	Unused	Unused
vCenter Server management network	N/A	N/A	50	Standby	Active	Unused	Unused
VxRail management network	N/A	vmk1	50	Standby	Active	Unused	Unused
vSAN network	100%	vmk3	200	Unused	Unused	Active	Standby
vMotion network	50%	vmk4	100	Unused	Unused	Standby	Active
Witness traffic	N/A	vmk5	60	Active	Standby	Unused	Unused
Virtual machines	60%	N/A	N/A	Active	Standby	Unused	Unused

Table 6.5 – The network port groups layout on the VxRail vSAN two-node cluster in Figure 6.10

A vSAN witness is a virtual appliance; it comes with two VMkernel network interfaces, vmk0 and vmk1. vmk0 is used for the management network traffic, and vmk1 is used for the vSAN network traffic. The VxRail node's vmk5 VMkernel network interface is tagged with *witness* network traffic. In *Figure 6.10*, the VxRail node must have a static route configured for vmk5 and be able to access vmk1 on the vSAN witness. vmk1 is tagged with *vSAN* network traffic. The vSAN witness also has a static route configured for vmk1 and can access vmk5 on the VxRail nodes.

Ensure some incoming and outgoing firewall ports for the following services are open, as shown in *Table 6.6*:

Services	Port	Protocol	To	From
vSAN Clustering Service	`12345, 23451`	UDP	VxRail nodes	VxRail nodes
vSAN Transport	`2233`	TCP	VxRail nodes	VxRail nodes
vSAN VASA Vendor Provider	`8080`	TCP	VxRail nodes and vCenter Server	VxRail nodes and vCenter Server
vSAN Unicast Agent to the vSAN Witness	`12321`	UDP	VxRail nodes and vSAN Witness	VxRail nodes and vSAN Witness

Table 6.6 – Service ports on VxRail appliance

> **Important Note**
>
> The vSAN Witness virtual appliance does not require the optional vSphere licenses. The vSphere license is hardcoded in the vSAN witness virtual appliance.

In the scenario in *Figure 6.10*, vCenter Server and the vSAN Witness are installed in the main data center. vCenter Server must be deployed before you deploy the VxRail vSAN two-node cluster. This vCenter Server version must be 6.7 Update 1 or later. The next section will discuss the external vCenter Server settings.

External vCenter Server

In the previous section, we already mentioned that VxRail vSAN two-node clusters only support the external vCenter Server. Before deploying a VxRail vSAN two-node cluster, you need to check the VxRail and external vCenter interoperability matrix in *Figure 6.11* and make sure the VxRail software edition can support the external vCenter Server. You can find more information at this link: `https://www.dell.com/support/kbdoc/en-us/000157682/vxrail-vxrail-and-external-vcenter-interoperability-matrix`.

In this interoperability matrix, you can verify the compatibility of a VxRail software release and the external vCenter Server version:

Matrix of supported external vCenter versions with VxRail

VxRail Release	ESXi (Version - Build #)	External vCenter versions	
		Minimum	Recommended
7.0.371	7.0 U3d - 19482537	7.0 U3 (7.0.3)	7.0 U3d (7.0.3.00500) or later
7.0.370	7.0 U3d - 19482537	7.0 U3 (7.0.3)	7.0 U3d (7.0.3.00500) or later
7.0.360 [30]	7.0 U3c - 19193900	7.0 U3 (7.0.3)	7.0 U3c (7.0.3.00300) or later
7.0.350	7.0 U3c - 19193900	7.0 U3 (7.0.3)	7.0 U3c (7.0.3.00300) or later
7.0.320 [29]	7.0 U3c - 19193900	7.0 U3 (7.0.3)	7.0 U3c (7.0.3.00300) or later
7.0.300	7.0 U3 - 18644231	7.0 U3 (7.0.3)	7.0 U3a (7.0.3.00100) or later
7.0.241 [28]	7.0 U2c - 18426014	7.0 U2 (7.0.2)	7.0 U2d (7.0.2.00500) or later
7.0.240	7.0 U2c - 18426014	7.0 U2 (7.0.2)	7.0 U2c (7.0.2.00400) or later
7.0.210 [24]	7.0 U2a - 17867351	7.0 U2 (7.0.2)	7.0 U2b (7.0.2.00200) or later
7.0.203 [26]	7.0 U2a - 17867351	7.0 U2 (7.0.2)	7.0 U2b (7.0.2.00200) or later
7.0.202	7.0 U2a - 17867351	7.0 U2 (7.0.2)	7.0 U2b (7.0.2.00200) or later
7.0.201 [23]	7.0 U2a - 17867351	7.0 U2 (7.0.2)	7.0 U2b (7.0.2.00200) or later
7.0.200	7.0 U2a - 17867351	7.0 U2 (7.0.2)	7.0 U2 (7.0.2) or later
7.0.132	7.0 U1d - 17551050	7.0 U1 (7.0.1)	7.0 U1c (7.0.1.00200) or later
7.0.131	7.0 U1d - 17551050	7.0 U1 (7.0.1)	7.0 U1c (7.0.1.00200) or later
7.0.130	7.0 U1c - 17325551	7.0 U1 (7.0.1)	7.0 U1c (7.0.1.00200) or later
7.0.101 [20]	7.0 U1b - 17168206	7.0 U1 (7.0.1)	7.0 U1 (7.0.1) or later
7.0.100 [20]	7.0 U1 - 16850804	7.0 U1 (7.0.1)	7.0 U1 (7.0.1) or later
7.0.010 [20]	7.0 b - 16324942	7.0 GA	7.0.0b or later
7.0.000 [20]	7.0 GA - 15843807	7.0 GA	7.0 GA or later

Figure 6.11 – VxRail and external vCenter interoperability matrix;
this information is copyright of Dell Technologies

Now, we will discuss a supported and unsupported configuration of the external vCenter Server on a VxRail vSAN two-node cluster.

Supported configuration

If VxRail comes with VxRail release 7.0.240, and the external vCenter Server version is 7.0 Update 2, according to the external vCenter interoperability matrix, it is a supported configuration.

Unsupported configuration

If the VxRail Appliance comes with VxRail release 7.0.320, and the external vCenter Server version is 7.0 Update 1c (7.0.1.00200), you should upgrade the external vCenter Server version to 7.0 U3c or later before deploying the VxRail vSAN two-node cluster; otherwise, it is an unsupported configuration.

> **Important Note**
>
> If you want more information on VMware vCenter build numbers and versions, you can find the details on this KB page: `https://kb.vmware.com/s/article/2143838`.

You have now learned about the network configuration, as well as software and hardware requirements for a VxRail vSAN two-node cluster. The next section will discuss failover scenarios of a VxRail vSAN two-node cluster.

Failure scenarios of a VxRail vSAN two-node cluster

This section will discuss some failure scenarios of a VxRail vSAN two-node cluster. The virtual machines allocated on the VxRail vSAN two-node cluster trigger different behavior when any hardware failure (for example, to the VxRail node, vSAN witness, HDD, or network uplinks) exists in the cluster. In *Figure 6.12*, there are four virtual machines (**VM A/B/C/D**) running on this VxRail vSAN two-node cluster. VMs A and B are allocated on **Fault Domain 1**, and VMs C and D are allocated on **Fault Domain 2**. A disk group (one SSD cache and three HDDs) is created on each VxRail node:

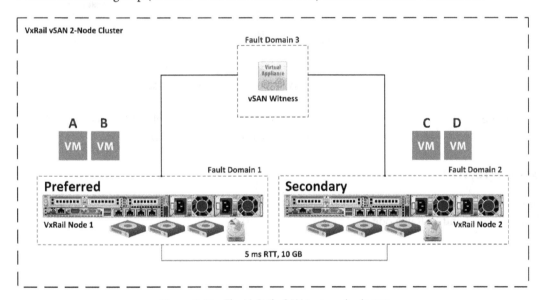

Figure 6.12 – The VxRail vSAN two-node cluster

The next section will discuss the four failover scenarios, including FD hardware failures, a faulty vSAN witness, disconnected network uplinks, and HDD failures.

Scenario one

In *Figure 6.13*, what status will the virtual machines trigger if VxRail Node 2 is faulty in the VxRail vSAN two-node cluster?

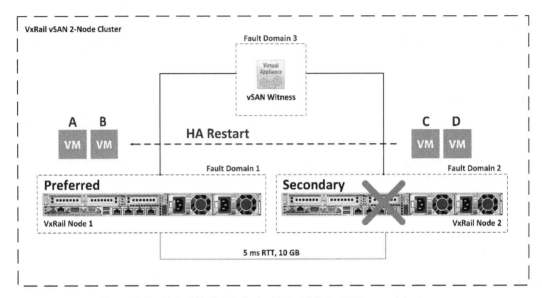

Figure 6.13 – VxRail Node 2 is faulty in the VxRail vSAN two-node cluster

VMs A and B keep running, but VMs C and D will shut down and trigger the **HA Restart** in VxRail Node 1. This is a normal hardware failure case; it will trigger the vSphere HA.

Scenario two

In *Figure 6.14*, what status will the **VMs** trigger if the vSAN communication is disconnected between VxRail Node 1 and VxRail Node 2?

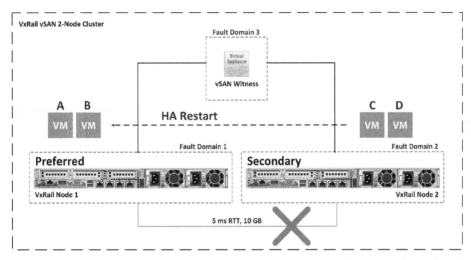

Figure 6.14 – The vSAN communication is disconnected between VxRail Node 1 and VxRail Node 2

VMs A and B keep running, but VMs C and D will shut down and trigger the HA Restart in VxRail Node 1. It is a split-brain scenario in a vSAN stretched cluster. When the data nodes (VxRail Node 1 and VxRail Node 2) cannot communicate and the vSAN witness is available, the VMs allocated on Fault Domain 2 (Secondary) will shut down once and restart in **Fault Domain 1** (Preferred).

Scenario three

In *Figure 6.15*, what status will the virtual machines trigger if the communication between a vSAN witness and two VxRail nodes is disconnected?

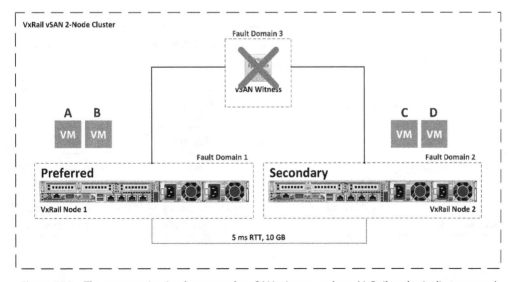

Figure 6.15 – The communication between the vSAN witness and two VxRail nodes is disconnected

All virtual machines (VM A/B/C/D) keep running in the VxRail vSAN two-node cluster because the communication of two VxRail nodes can be connected.

Scenario four

In *Figure 6.16*, what status will the virtual machines trigger if one HDD is faulty in the disk group on VxRail Node 1?

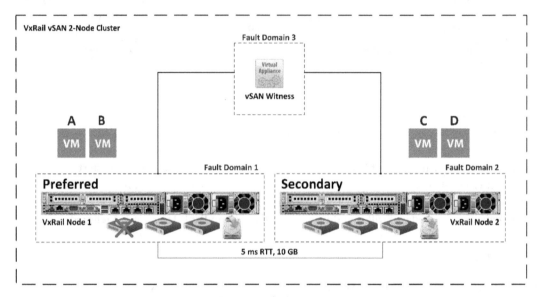

Figure 6.16 – The HDD is faulted on VxRail Node 1

All virtual machines (VM A/B/C/D) keep running in the VxRail vSAN two-node cluster, but some vSAN objects are degraded in VxRail Node 1. The data rebuild process will be started in the disk group after the faulty HDD is replaced on VxRail Node 1.

With the preceding failure scenarios, you learned about the different behaviors of virtual machines. *Table 6.7* shows a summary of the expected results for the different failure scenarios:

Failure scenarios	The expected result
One of the nodes is faulty in the VxRail vSAN two-node cluster.	The virtual machines allocated to the faulty node will shut down and restart in the standby node.
The communication of two VxRail nodes is disconnected.	The virtual machines allocated to the secondary FD will shut down and restart in the preferred FD.

The communication between two VxRail nodes and the vSAN witness is disconnected.	All virtual machines allocated on the preferred and secondary FDs keep running.
The HDD is faulty on one of the VxRail nodes.	Some of the vSAN objects are degraded in the disk group.

Table 6.7 – The expected results for the different failure scenarios

After going through the four failure scenarios, you now understand the expected result in each scenario.

Summary

In this chapter, we provided an overview and learned about the design of a VxRail vSAN two-node cluster, including the network, hardware, and software requirements and some failure scenarios. When you plan to design a disaster recovery solution or a small-size VMware environment, the VxRail vSAN two-node cluster is a good option for you.

In the next chapter, you will learn about the design of a stretched cluster on VxRail and the best practices of this solution.

Questions

The following are a short list of review questions to help reinforce your learning and help you identify areas which require some improvement.

1. Which deployment options can be supported on a VxRail vSAN two-node cluster?

    A. Central management and localized witness

    B. Central management

    C. Central management and witness

    D. Central witness

    E. Localized management and witness

    F. Localized witness

    G. All of the above

2. Which two network configurations can be supported on a VxRail vSAN two-node cluster?

   A. Direct-connect configuration

   B. Fibre Channel configuration

   C. iSCSI channel configuration

   D. Switch configuration

   E. 10/25 Gb network configuration

   F. None of the above

3. What are the minimum network ports required on a VxRail vSAN two-node cluster?

   A. Two 10 Gb ports

   B. Three 10 Gb ports

   C. Four 1 Gb ports

   D. Four 10/25 Gb ports

   E. Five 10 Gb ports

   F. Five 25 Gb ports

4. Which network configurations can be supported on a VxRail vSAN two-node cluster?

   A. Two 1 Gb ports and two 10 Gb ports

   B. Four 1 Gb Gb ports

   C. Two 10 Gb ports and two 25 Gb ports

   D. Four 10 Gb ports

   E. Four 25 Gb ports

   F. All of the above

5. Which VxRail software releases can support switch configuration on a VxRail vSAN two-node cluster?

   A. VxRail 4.7.200

   B. VxRail 4.7.300

   C. VxRail 4.7.410

   D. VxRail 7.0.240

   E. VxRail 7.0.300

   F. All of the above

6. Which VxRail Series can support a VxRail vSAN two-node cluster?

    A. VxRail E-Series

    B. VxRail P-Series

    C. VxRail V-Series

    D. VxRail D-Series

    E. VxRail S-Series

    F. All of the above

7. What is the maximum supported RTT between a VxRail vSAN two-node cluster and vSAN witness?

    A. 100 milliseconds

    B. 200 milliseconds

    C. 300 milliseconds

    D. 400 milliseconds

    E. 500 milliseconds

    F. None of the above

8. Which description is true for the VxRail vSAN two-node cluster deployment central management and localized witness?

    A. It can provide a single management dashboard to manage all VxRail vSAN two-node clusters.

    B. Extra network configuration and expenditure are not required for the communication of vCenter Server and the vSAN witness.

    C. Optional vCenter Server licenses and hardware are not required.

    D. Extra software and hardware costs are required for the deployment of multiple vCenter Server vSAN witnesses, for example, an optional vCenter Server license per site.

    E. It cannot provide a single management dashboard to manage all VxRail vSAN two-node clusters.

9. Which description is true for the VxRail vSAN two-node cluster deployment localized management and witness?

    A. It can provide a single management dashboard to manage all VxRail vSAN two-node clusters.

    B. Extra network configuration and cost are not required for the communication of vCenter Server and the vSAN witness.

    C. Optional vCenter Server licenses and hardware are not required.

D. Extra software and hardware costs are required for the deployment of multiple vCenter Server vSAN witnesses, for example, an optional vCenter Server license per site.

E. It cannot provide a single management dashboard to manage all VxRail vSAN two-node clusters.

10. Which description is true for the VxRail vSAN two-node cluster deployment central management and witness?

A. It can provide a single management dashboard to manage all VxRail vSAN two-node clusters.

B. Extra network configuration and expenditure are not required for the communication of vCenter Server and the vSAN witness.

C. Optional vCenter Server licenses and hardware are not required.

D. Extra software and hardware costs are required for the deployment of multiple vCenter Server vSAN witnesses, for example, an optional vCenter Server license per site.

E. It cannot provide a single management dashboard to manage all VxRail vSAN two-node clusters.

11. What network port group is configured as 100% NIOC shared on vDS?

A. Management network

B. vSAN network

C. vMotion network

D. Virtual machines

E. vCenter Server management network

F. VxRail management network

12. What behavior will the virtual machines trigger if the vSAN communication is disconnected between two VxRail nodes?

A. The virtual machines keep running.

B. All of the virtual machines shut down.

C. The virtual machines on the preferred FD will shut down and power on the secondary FD.

D. The virtual machines on the secondary FD will shut down and power on the preferred FD.

E. None of the above.

Part 3: Design of Data Protection for the VxRail System

VxRail can provide high-availability solutions. The reader will learn about the design of advanced solutions for VxRail – for example, stretched clusters on VxRail, Dell EMC RecoverPoint for VMs, and VxRail with Veeam Backup & Replication.

This part of the book comprises the following chapters:

7
Design of Stretched Cluster on VxRail

In the previous chapter, you learned about the design of the VxRail vSAN Standard and two-node cluster, including the network, hardware, and software requirements, and some failure scenarios. When you plan to design a disaster recovery solution or a small-sized VMware environment, the VxRail vSAN two-node cluster is a great choice.

Disaster recovery is a critical factor for all systems and virtual infrastructures. VxRail Appliance delivers different disaster recovery solutions. VxRail Stretched Cluster is one of these options; it can provide the synchronous replication of data across two sites located at separate geographical locations. This solution allows the management of the failure of an entire site and supports the different **Failure Tolerance Methods (FTMs)**. In this chapter, you will see an overview of VxRail Stretched Cluster, the planning and designing of VxRail Stretched Cluster, and the different failure scenarios.

This chapter includes the following main topics:

- Overview of VxRail Stretched Cluster
- Design of VxRail Stretched Cluster
- A scenario using VxRail Stretched Cluster
- Failure scenarios of VxRail Stretched Cluster

Overview of VxRail Stretched Cluster

VxRail Stretched Cluster is designed for an active-active data center solution, and it can provide redundancy and failure protection across two separate physical sites. In *Figure 7.1*, VxRail Stretched Cluster is built between two individual sites, including two data sites (**Preferred Site** and **Secondary Site**) and one witness site. The witness host deploys a third site that contains the witness components of the VM objects. The VMs can continue to provide the services when one of the data sites fails. This is the main feature of VxRail Stretched Cluster:

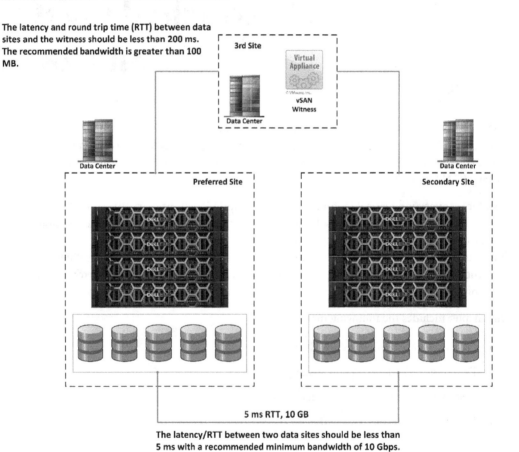

Figure 7.1 – VxRail Stretched Cluster with eight nodes

If you choose the deployment of VxRail Stretched Cluster, you need to consider the following:

- The minimum supported configuration for VxRail Stretched Cluster is 1+1+1 (two VxRail nodes plus one witness). The maximum supported configuration is 15+15+1 (30 VxRail nodes plus 1 witness).

- VxRail Stretched Cluster must be deployed across two individual physical sites in an active-active configuration.

- Starting from VxRail version 4.7.0, the configurations of 2+2+1 (four VxRail nodes plus one witness) are supported.

- Starting from VxRail version 7.0, either a VxRail vCenter Server or a customer-supplied vCenter Server instance can be used for VxRail Stretched Cluster, but a customer-supplied vCenter Server instance is recommended.

- The vSAN witness appliance or a physical ESXi host can be used as a witness. The vSAN witness appliance includes a vSphere license. The physical host requires a vSphere license.

- The vSAN witness cannot be a part of VxRail Stretched Cluster. Also, it must have one VMkernel adapter with vSAN traffic enabled and connected to all nodes in VxRail Stretched Cluster. The witness uses a separate VMkernel adapter for managing traffic.

- Both Layer 2 and Layer 3 support the vSAN connectivity between the data nodes across two sites. A static route is required if it is Layer 2, but in Layer 3, it is not required.

- The maximum supported **Round Trip Time (RTT)** between the data nodes and vSAN witness is 200 milliseconds. For configurations up to 10+10+1, latency or RTT less than or equal to 200 milliseconds is acceptable. For configuration greater than 10+10+1, latency or RTT less than or equal to 100 milliseconds is required.

- The maximum supported RTT among the data nodes is 5 milliseconds.

- A bandwidth of 10 GB between the data sites is recommended.

- A bandwidth of 100 Mb between the data sites and the witness site is recommended.

Next, let's look at its architecture.

The architecture of VxRail Stretched Cluster

The architecture of VxRail Stretched Cluster includes the data nodes and one witness node. In *Figure 7.2*, each node is configured as a **Fault Domain** (**FD**); VxRail nodes **1**, **2**, **3**, and **4** are configured to **Fault Domain 1** (preferred), VxRail nodes **5**, **6**, **7**, and **8** are configured to **Fault Domain 2** (secondary), and **vSAN Witness** is configured to **Fault Domain 3**. It supports three types of site disaster tolerance— dual-site mirroring, keeping data in the preferred site, and keeping data in the secondary site:

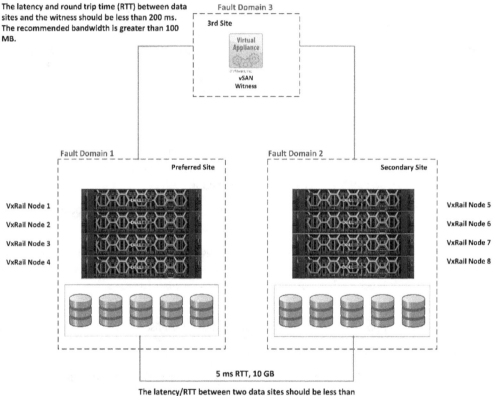

Figure 7.2 – Three FDs of VxRail Stretched Cluster

When three FDs are available and the **Failure to Tolerate** (**FTT**) parameter is set to RAID-1, a VM is created on VxRail Stretched Cluster that has mirror protection, one copy of the data on FD 1 and the second copy of the data on FD 2. The witness component is placed on the vSAN witness.

The next section will discuss the types of site disaster tolerance for VxRail Stretched Cluster.

Site disaster tolerance

When you create a storage policy for VxRail Stretched Cluster, two parameters are required: **site disaster tolerance** and **FTT**. For VxRail Stretched Cluster, you can choose the following options from the **Site disaster tolerance** menu:

- **Site mirroring - stretched cluster**: This enables data protection across the preferred and secondary sites
- **None - keep data on Preferred (stretched cluster)**: This keeps the data on the preferred site only and there is no cross-site protection
- **None - keep data on Secondary (stretched cluster)**: This keeps the data on the secondary site only and there is no cross-site protection

If you do not deploy VxRail Stretched Cluster, you can choose the following options from the **Site disaster tolerance** menu:

- **None - standard cluster**: This is used to deploy a standalone VxRail cluster
- **Host mirroring - 2 node cluster**: This is used to deploy the VxRail vSAN two-node cluster
- **None - stretched cluster**: This is used to deploy the non-stretched cluster

You can see an overview of these parameters in *Figure 7.3*:

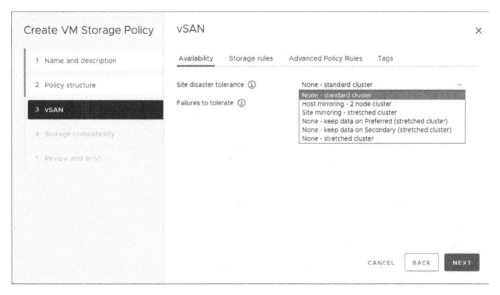

Figure 7.3 – Site disaster tolerance parameters

The next section will discuss the **Failures to tolerate** options for VxRail Stretched Cluster.

Failures to tolerate

VxRail Stretched Cluster can support different types of protection; these parameters are used to define the protection methodologies. You can select the following options from the **Failures to tolerate** menu:

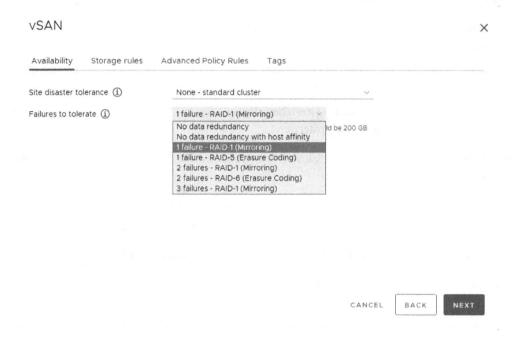

Figure 7.4 – Failures to tolerate parameters

Here are brief descriptions of what these options mean:

- **No data redundancy**: No data protection in both the preferred and secondary sites.

- **No data redundancy with host affinity**: No data protection in the preferred site or secondary site, and it is controlled by host affinity rules.

- **1 failure - RAID-1 (Mirroring)**: It is a mirrored volume and gives good performance across the preferred and secondary sites, and full redundancy with 200% capacity usage.

- **1 failure - RAID-5 (Erasure Coding)**: It is a stripped volume with parity. It has good performance with redundancy that requires a minimum of five nodes per site.

- **2 failures - RAID-1 (Mirroring)**: It is a mirrored volume and there is good performance across the preferred and secondary sites, and full redundancy with 600% capacity usage.

- **2 failures - RAID-6 (Erasure Coding)**: It is a stripped volume with double parity. It has good performance with redundancy that requires a minimum of seven nodes per site.

- **3 failures - RAID-1 (Mirroring)**: It is a mirrored volume and there is good performance across the preferred and secondary sites, and full redundancy with 800% capacity usage.

The following sections will discuss sample configurations of a VM storage policy in VxRail Stretched Cluster, including non-dual-site mirroring, RAID-1, and RAID-5 configurations.

Site mirroring with RAID-1

Figure 7.5 shows VxRail Stretched Cluster with eight nodes; nodes **1**, **2**, **3**, and **4** are installed in the preferred site and nodes **5**, **6**, **7**, and **8** are installed in the secondary site. One VM (100 GB storage) is running in the preferred site; the VM storage policy with the following parameters is found in this VM:

- **Site Disaster Tolerance: Site mirroring - stretched cluster**

- **Failures To Tolerate: 1 failure - RAID-1 (Mirroring)**

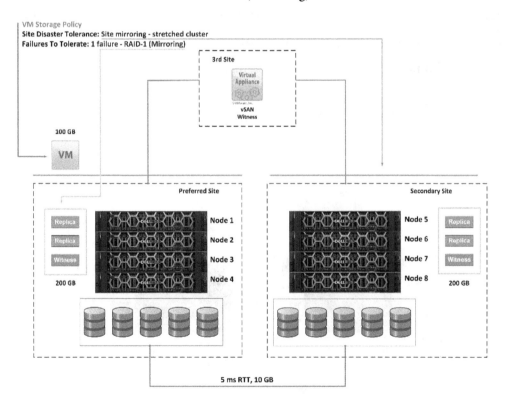

Figure 7.5 – Site mirroring with RAID-1 in VxRail Stretched Cluster

In the example of *Figure 7.5*, due to the **Site Disaster Tolerance** parameter being set to **Site mirroring - stretched cluster**, the VM can be protected by RAID-1 mirroring across the preferred and secondary sites. The **Failures to tolerate** parameter is set to **1 failure - RAID-1 (Mirroring)**, so the VM can also be protected by RAID-1 mirroring in the preferred and secondary sites.

Using this VM storage policy setting, the VM has a full copy of **Replica** if one of the sites fails. Also, the VM has a full data copy if one of the four nodes fails in each site (preferred and secondary sites). If you choose this protection option, it can deliver site resilience and mirroring protection per site, but it requires four times more storage requirements than the protected VM. In this example, the storage capacity of the VM is 100 GB, and it requires 400 GB storage capacity in VxRail Stretched Cluster if you apply this storage policy to the VM. In the next section, we will discuss site mirroring with RAID-5 protection in VxRail Stretched Cluster across two sites.

Site mirroring with RAID-5

Figure 7.6 shows VxRail Stretched Cluster with eight nodes; nodes **1**, **2**, **3**, and **4** are installed in the preferred site and nodes **5**, **6**, **7**, and **8** are installed in the secondary site. One VM (100 GB storage) is running in the preferred site; the VM storage policy with the following parameters is found in this VM:

- **Site Disaster Tolerance: Site mirroring - stretched cluster**
- **Failures To Tolerate: 1 failure - RAID-5 (Erasure Coding)**

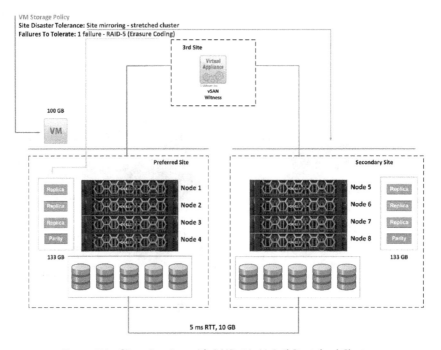

Figure 7.6 – Site mirroring with RAID-5 in VxRail Stretched Cluster

In this example, due to the **Site Disaster Tolerance** parameter being set to **Site mirroring - stretched cluster**, the VM can be protected by RAID-1 mirroring across the preferred and secondary sites. If the **Failures To Tolerate** parameter is set to **1 failure - RAID-5 (Erasure Coding)**, the VM can be protected by RAID-5 protection in the preferred and secondary sites.

Using these **VM Storage Policy** settings, the VM has a full copy of **Replica** if one of the sites fails. The VM also has RAID-5 protection if one of the four nodes fails in each site (preferred and secondary sites). If you choose this protection option, it can deliver site resilience and RAID-5 protection per site, but it requires 2.66 times the protected VM's storage requirements. In this example, the storage capacity of the VM is 100 GB, and it requires 266 GB storage capacity in VxRail Stretched Cluster if you apply this VM storage policy to the VM. Compared to the example in *Figure 7.5*, the storage capacity requirement is lower than RAID-1 mirroring. In the next section, we will discuss site affinity rules in VxRail Stretched Cluster across two sites.

> **Important note**
>
> VxRail All-Flash Stretched Cluster only supports **RAID-5 (Erasure Coding)** and **RAID-6 (Erasure Coding)**.

Keeping the data in the preferred site

Figure 7.7 shows VxRail Stretched Cluster with eight nodes; nodes **1**, **2**, **3**, and **4** are installed in the preferred site and nodes **5**, **6**, **7**, and **8** are installed in the secondary site. One VM (100 GB storage) is running in the preferred site; the VM storage policy with the following parameters is found in this VM:

- **Site Disaster Tolerance: None - keep data on Preferred (stretched cluster)**
- **Failures To Tolerate: 1 failure - RAID-1 (Mirroring)**

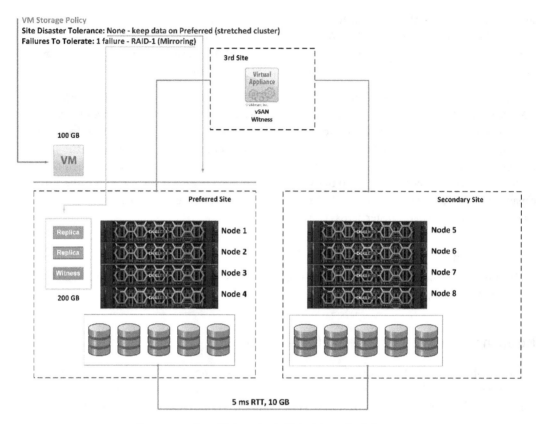

Figure 7.7 – Site affinity rules in VxRail Stretched Cluster

In this example, due to the **Site Disaster Tolerance** parameter being set to **None - keep data on Preferred (stretched cluster)**, all VMs are only stored in the preferred site. Also, the **Failures To Tolerate** parameter is set to **1 failure - RAID-1 (Mirroring)**, so the VM can also be protected by RAID-1 in the preferred site.

Using these VM storage policy settings, the VM has a full data copy stored in the nodes in the preferred site. The VM also has RAID-1 protection if one of four nodes fails in the preferred site. If you choose this protection option, it *cannot* deliver site resilience and only provides RAID-1 protection in one site based on affinity rule settings, but it requires twice the amount of storage requirements as the protected VM. In this example, the storage capacity of the VM is 100 GB, and it requires 200 GB storage capacity in VxRail Stretched Cluster if you apply this VM storage policy to the VM. Compared to the examples in *Figure 7.5* and *Figure 7.6*, the storage capacity requirement is lower in this scenario. The preceding examples help you understand VM storage policies in VxRail Stretched Cluster.

Table 7.1 shows a summary of all VM storage policy settings in VxRail Stretched Cluster, assuming the VM's storage capacity is 100 GB:

Site Availability	Capacity Requirement in Preferred Site	Capacity Requirement in Secondary Site	Capacity Requirement
Dual-site mirroring without redundancy	100 GB	100 GB	200 GB (2x)
Dual-site mirroring with RAID-1 (one failure)	200 GB	200 GB	400 GB (4x)
Dual-site mirroring with RAID-1 (two failures)	300 GB	300 GB	600 GB (6x)
Dual-site mirroring with RAID-1 (three failures)	400 GB	400 GB	800 GB (8x)
Dual-site mirroring with RAID-5 (one failure)	133 GB	133 GB	266 GB (2.66x)
Dual-site mirroring with RAID-6 (two failures)	150 GB	150 GB	300 GB (3x)
Preferred site only with RAID-1 (one failure)	200 GB	0	200 GB (2x)
Preferred site only with RAID-1 (two failures)	300 GB	0	300 GB (3x)
Preferred site only with RAID-1 (three failures)	400 GB	0	400 GB (4x)
Preferred site only with RAID-5 (one failure)	133 GB	0	133 GB (1.3x)
Preferred site only with RAID-6 (two failures)	150 GB	0	150 GB (1.5x)
Secondary site only with RAID-1 (one failure)	0	200 GB	200 GB (2x)
Secondary site only with RAID-1 (two failures)	0	300 GB	300 GB (3x)
Secondary site only with RAID-1 (three failures)	0	400 GB	400 GB (4x)
Secondary site only with RAID-5 (one failure)	0	133 GB	133 GB (1.3x)
Secondary site only with RAID-6 (two failures)	0	150 GB	150 GB (1.5x)

Table 7.1 – Summary of all VM storage policy settings in VxRail Stretched Cluster

The following section will discuss the design of VxRail Stretched Cluster, including **Witness Traffic Separation (WTS)** and vCenter Server options.

Design of VxRail Stretched Cluster

When you design VxRail Stretched Cluster, network infrastructure is an important factor for its deployment. This section will discuss WTS, which is an optional network configuration. We will discuss VxRail Stretched Cluster with and without WTS. This feature provides a flexible network configuration by separating the network of data nodes into data nodes and then data nodes to witness traffic. All operational tasks of VxRail Stretched Cluster are delivered with a vCenter Server instance; it supports two deployment options. The following sections will discuss the design of WTS and the VMware vCenter Server deployment options.

VxRail Stretched Cluster without WTS

Figure 7.8 shows a standard VxRail Stretched Cluster configuration without WTS. There are four data nodes installed in the preferred and secondary sites. The data nodes are connected through **Stretched Layer 2 vSAN Network** across the preferred and secondary sites. The vSAN communication between the preferred and secondary sites is routed over Layer 3 to the third site. If you are using this configuration, the data nodes use a static route from the vSAN VMkernel interface to the vSAN witness VMkernel interface tagged from the vSAN traffic:

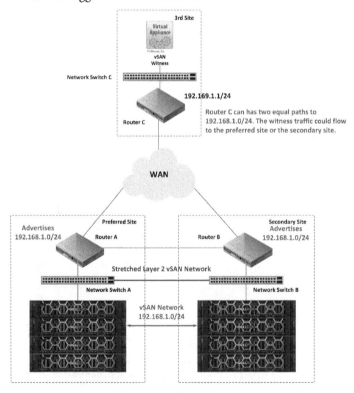

Figure 7.8 – VxRail Stretched Cluster without WTS

In this configuration, the witness traffic cannot be separated from the vSAN traffic. The witness traffic can flow to either the preferred or secondary site through the `192.168.1.0/24` network. This configuration is not recommended. **Router C** in the third site has two equally costing paths to the `192.168.1.0/24` network, and the network problem will happen when the traffic is sent back from the third site. It can cause issues if stateful firewall rules exist on each site. VxRail Stretched Cluster with WTS is recommended, which will be discussed in the following section.

VxRail Stretched Cluster with WTS

Now, we will discuss VxRail Stretched Cluster with WTS. *Figure 7.9* shows a standard VxRail Stretched Cluster configuration with WTS:

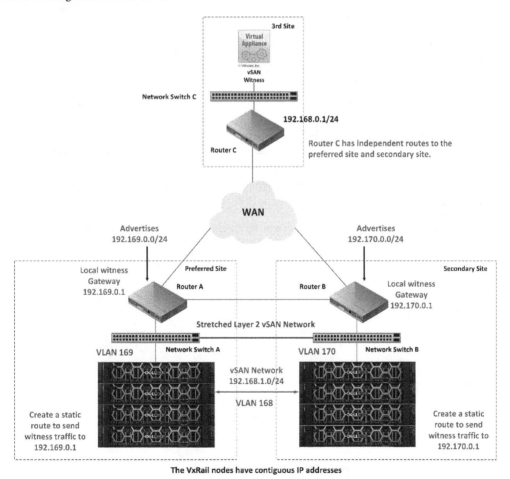

Figure 7.9 – VxRail Stretched Cluster with WTS

In this configuration, you need to create a distributed port group tagged as **VLAN 169** in the preferred site and a port group tagged as **VLAN 170** in the secondary site. Also, create a new witness VMkernel interface and add it to each data node in each site. The local witness gateway is used at each site, which is used to route the witness traffic from the data nodes to the vSAN witness. The preferred site advertises the 192.169.0.0/24 network to the third site, and the secondary site advertises the 192.170.0.0/24 network to the third site. Now, you can create static routes between the data nodes to send and receive witness traffic from the data nodes to the vSAN witness. If you use this configuration, the witness traffic is on a separate VLAN from the vSAN network because the third site has two independent preferred and secondary site routes, and you can enable the contiguous vSAN IP addresses on each data node across preferred and secondary sites.

With the preceding two examples, you will now understand the difference between VxRail Stretched Cluster with WTS and without WTS.

The following section will discuss the design of the embedded vCenter Server instance and customer-supplied vCenter Server instance.

VMware vCenter Server options

During the deployment of VxRail Stretched Cluster, we have two options for the vCenter Server instance—that is, embedded vCenter Server (internal vCenter Server) and customer-supplied vCenter Server (external vCenter Server). An embedded vCenter Server is deployed on the VxRail cluster during VxRail's initial installation and a customer-supplied vCenter Server is deployed in your VMware environment. Customer-supplied vCenter Server is the recommended configuration.

In *Figure 7.10*, VxRail Stretched Cluster is managed by an embedded vCenter Server instance. In this deployment, the embedded vCenter Server instance can only manage a VxRail cluster—for example, a VxRail standard cluster, VxRail vSAN two-node cluster, or VxRail Stretched Cluster. If you are using this deployment, the lifecycle management is also supported by VxRail Stretched Cluster. Starting with VxRail 7.0, either an embedded vCenter Server or a customer-supplied vCenter Server can be used for VxRail Stretched Cluster.

Figure 7.10 shows VxRail Stretched Cluster with an embedded vCenter Server instance:

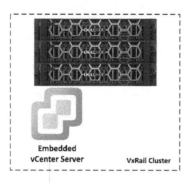

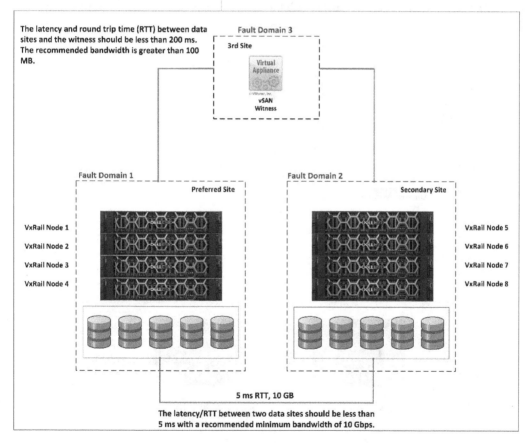

Figure 7.10 – VxRail Stretched Cluster with an embedded vCenter Server instance

In *Figure 7.11*, VxRail Stretched Cluster is managed by a customer-supplied vCenter Server instance. In this deployment, the customer-supplied vCenter Server instance can manage both a VxRail cluster and a non-VxRail cluster—for example, a VxRail standard cluster, a vSphere cluster, or a VxRail Stretched Cluster. If you are using this deployment, it has the following advantages:

- It can provide central management of VxRail clusters and non-VxRail clusters.

- You can easily migrate the VM's workload between the VxRail cluster and the non-VxRail cluster.

- It can manage different types of VxRail clusters—that is, a VxRail All-Flash cluster and a hybrid cluster.

- Each VxRail cluster with a customer-supplied vCenter Server instance has its own VxRail Manager.

Figure 7.11 shows VxRail Stretched Cluster with the customer-supplied vCenter Server instance:

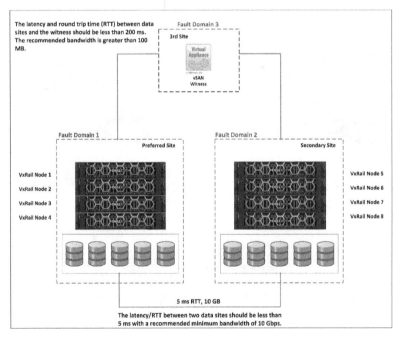

Figure 7.11 – VxRail Stretched Cluster with the customer-supplied vCenter Server instance

With *Figure 7.10* and *Figure 7.11*, you now understand the differences between both vCenter Server instances. *Table 7.2* shows a comparison of vCenter features:

	Embedded vCenter Server	**Customer-supplied vCenter Server**
vCenter Server License	The vCenter Server Standard license is bundled with VxRail Appliance.	It requires the optional vCenter Server Standard license.
Compatibility	It is bundled with VxRail software that is compatible with VxRail Manager and vSphere software.	You need to verify the VxRail and external vCenter interoperability matrix.
Deployment	It will deploy automatically into the VxRail cluster during VxRail initialization.	You need to deploy it manually into your VMware environment.
Lifecycle management	VxRail's one-click upgrade includes this feature.	It does not include this feature.
vCenter migration	It cannot be migrated from the VxRail cluster.	It can be migrated to either a VxRail cluster or a non-VxRail cluster.
Cluster migration	It can be migrated to a customer-supplied vCenter Server.	It cannot be migrated to an embedded vCenter Server.
Multihoming	It does not support multihoming.	It can support multihoming.
Single sign-on domain	It only supports a single `vsphere.local` single sign-on (SSO) domain.	It has no domain restrictions.

Table 7.2 – vCenter feature comparison

When designing VxRail Stretched Cluster, you can choose which configuration is suitable for your environment. The next section will discuss a sample configuration of VxRail Stretched Cluster.

A scenario using VxRail Stretched Cluster

This section will discuss a sample configuration of VxRail Stretched Cluster, shown in *Figure 7.12*. Now, we will discuss VxRail Stretched Cluster with eight nodes, including the requirements, the design of the network configuration, and the VM storage policy configuration. When you design the deployment of VxRail Stretched Cluster, you need to consider all hardware and software requirements; you can refer to the *Overview of VxRail Stretched Cluster* section in this chapter for more details:

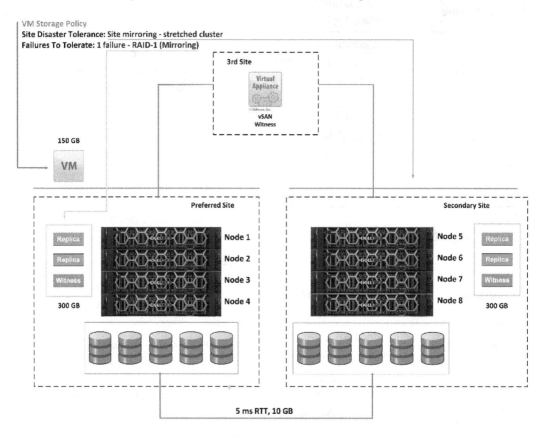

Figure 7.12 – VxRail Stretched Cluster with eight nodes

This configuration is a standard VxRail Stretched Cluster with eight nodes. *Table 7.3* shows the hardware configuration of each VxRail node. Each P670F model installed one 800 GB SSD (cache tier), four 7.68 TB SSDs (capacity tier), and one quad-port 10 GB Ethernet adapter:

VxRail model	VxRail P670F
CPU model	2 x Intel Xeon Gold 5317 3G, 12C/24T
Memory	512 GB (8 x 64 GB)
Network adapter	Intel Ethernet X710 Quad Port 10GbE SFP+, OCP NIC 3
Cache drive	1 x 800 GB SSD SAS drive
Capacity drive	4 x 7.68 TB SSD SAS drive

Table 7.3 – A sample configuration of VxRail P670F

Each VxRail P670F model includes the following software; *Table 7.4* shows the software edition of each VxRail and VMware component:

VxRail software release	7.0.370 build 27485531
VMware ESXi edition	7.0 Update 3d
VMware vCenter Server	7.0 Update 3d
VMware vSAN	7.0 Update 3d
VMware vSAN witness	7.0 Update 3d

Table 7.4 – VxRail software releases

The next section will discuss the network design of the scenario in *Figure 7.12*.

Network settings

Each VxRail P670F model has four 10 GB network ports, as shown in *Figure 7.13*. The **P1** and **P2** ports are used for ESXi with a VxRail management network and witness traffic. The **P3** and **P4** ports are used for the vSAN and vMotion networks:

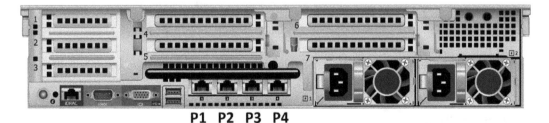

P1 P2 P3 P4

Intel Ethernet X710 Quad Port 10GbE

Figure 7.13 – Rear view of VxRail P670F

For the network design of VxRail Stretched Cluster, you can refer to the following table. *Table 7.5* shows a network layout used for VxRail Stretched Cluster with WTS:

Network Traffic	NIOC Shares	VMkernel ports	VLAN	P1	P2	P3	P4
Management network	40%	vmk2	101	Standby	Active	Unused	Unused
vCenter Server management network	N/A	N/A	101	Standby	Active	Unused	Unused
VxRail management network	N/A	vmk1	101	Standby	Active	Unused	Unused
vSAN network	100%	vmk3	200	Unused	Unused	Active	Standby
vMotion network	50%	vmk4	100	Unused	Unused	Standby	Active
Witness traffic	N/A	vmk5	102	Active	Standby	Unused	Unused
Virtual machines	60%	N/A	N/A	Active	Standby	Unused	Unused

Table 7.5 – Network layout of VxRail Stretched Cluster with WTS

Table 7.5 helps you understand VxRail Stretched Cluster's network layout. The next section will discuss the storage design of the scenario in *Figure 7.12*.

Storage settings

The VxRail P670F model installed one 800 GB SSD and four 7.68 TB SSDs; you can create a vSAN disk group with one 800 GB SSD for the cache tier and four 7.68TB SSDs for the capacity tier. For the disk groups upgrade, you can refer to the *Design of disk groups on VxRail P-Series* section in

Chapter 5. Since P670F is an All-Flash model, it can support RAID-1, RAID-5, and RAID-6 site protection. In the VM storage policy, make sure the **Site Disaster Tolerance** parameter is configured to **Site mirroring - stretched cluster** and the **Failures to tolerate** parameter is configured to **1 failure - RAID-1 (Mirroring)**. For the other type of site protection, you can refer to the *Overview of VxRail Stretched Cluster* section in this chapter. The next section will discuss the required software licenses for this scenario.

Software licenses

When you deploy VxRail Stretched Cluster, it requires the following VMware licenses. *Table 7.6* shows a summary of all the bundled software licenses on VxRail P670F:

Software Name	License Edition	Quantity	Remark
VMware vSphere	VMware vSphere Enterprise Plus per CPU	8	N/A
VMware vSAN	VMware vSAN Enterprise per CPU	8	You can choose the vSAN Enterprise or Enterprise Plus edition.
VMware vCenter Server	VMware vCenter Server Standard instance	1	Suggest deploying the customer-supplied vCenter Server. It requires an optional vCenter Server Standard license.
VMware vRealize Log Insight	Bundled with VxRail Appliance	1	N/A
VMware Replication	Bundled with VxRail Appliance	N/A	N/A
Dell EMC RecoverPoint for Virtual Machines	Bundled with VxRail Appliance	N/A	Five VM licenses per node

Table 7.6 – Summary of all bundled software licenses on VxRail P670F

With the preceding information, you understand the requirements of VxRail Stretched Cluster in *Figure 7.12*.

Failure scenarios of VxRail Stretched Cluster

This section will discuss some failure scenarios of VxRail Stretched Cluster. The VMs allocated on VxRail Stretched Cluster trigger different behavior when any hardware failure (for example, VxRail node, vSAN witness, HDD, network uplinks, and so on) exists in the cluster. In *Figure 7.14*, two VMs (**VM A** and **VM B**) are running on this VxRail Stretched Cluster instance; these two VMs are allocated on the preferred site, and the different VM storage policies assign each VM. **VM A** is configured with **VM Storage Policy A**, and **VM B** is configured with **VM Storage Policy B**:

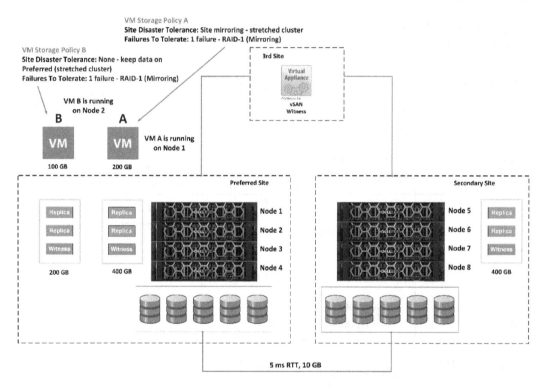

Figure 7.14 – VxRail Stretched Cluster with eight nodes

Now, we will discuss each failure scenario.

Failure scenario one

In *Figure 7.15*, what status will the VMs trigger if the vSAN communication is disconnected between the preferred and secondary sites?

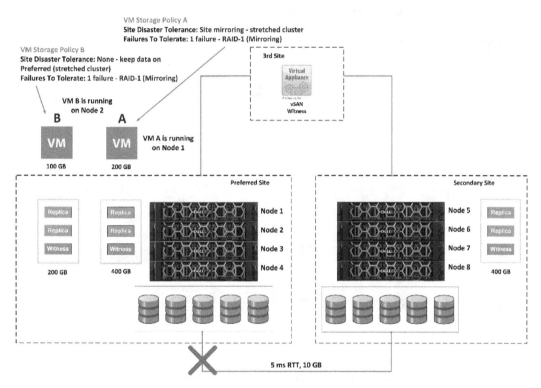

Figure 7.15 – Failure scenario one of VxRail Stretched Cluster

VM A and **VM B** keep running in this scenario, but the vSAN object status of **VM A** will be degraded because **VM Storage Policy A** is configured to **Site mirroring - stretched cluster**. The vSAN object status of **VM B** is healthy because **VM Storage Policy B** is configured to **None - keep data on Preferred (stretched cluster)**. When the data nodes (preferred and secondary sites) cannot communicate and the vSAN witness is available, the VMs allocated on the secondary site will shut down and restart in the preferred site.

Failure scenario two

In *Figure 7.16*, what status will the VMs trigger if **Node 1** fails in VxRail Stretched Cluster?

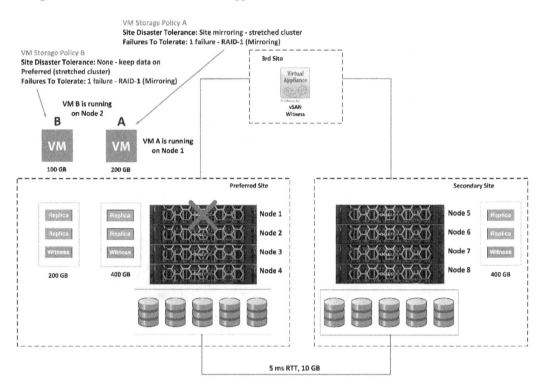

Figure 7.16 – Failure scenario two of VxRail Stretched Cluster

VM B keeps running, but the vSAN object status of **VM B** will be degraded, and one of the replicas will be built into **Node 4** because **VM Storage Policy B** is configured to **1 failure - RAID-1 (Mirroring)**. **VM A** will shut down and trigger the high-availability restart into **VxRail Node 4**, and one of the replicas will be built in **Node 4**. This is a normal hardware failure case, triggering vSphere High Availability.

Failure scenario three

In *Figure 7.17*, what status will the VMs trigger if the communication of the vSAN witness and VxRail's data nodes is disconnected?

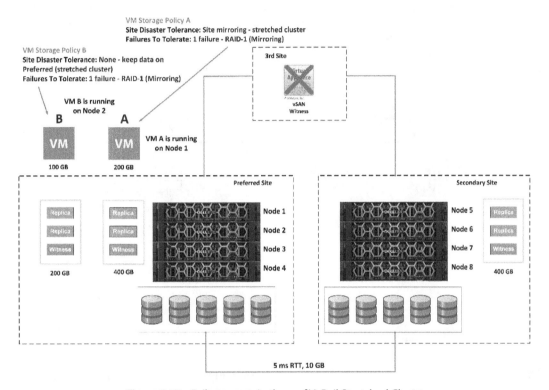

Figure 7.17 – Failure scenario three of VxRail Stretched Cluster

All VMs (**VM A** and **VM B**) keep running in VxRail Stretched Cluster because the communication between two VxRail nodes can be connected.

Failure scenario four

In *Figure 7.18*, what status will the VMs trigger if the preferred site fails in VxRail Stretched Cluster?

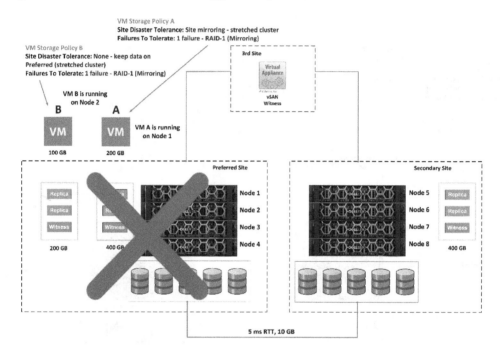

Figure 7.18 – Failure scenario four of VxRail Stretched Cluster

VM A will shut down and trigger the high-availability restart into the VxRail nodes in the secondary site. The vSAN object status of **VM A** will be degraded because **VM Storage Policy A** is configured to **Site mirroring - stretched cluster**. This is a normal hardware failure case, triggering vSphere High Availability. **VM B** will shut down and cannot trigger the high-availability feature because **VM Storage Policy B** is configured to **None - keep data on Preferred (stretched cluster)**.

Failure scenario five

In *Figure 7.19*, what status will the VMs trigger if the secondary site fails in VxRail Stretched Cluster?

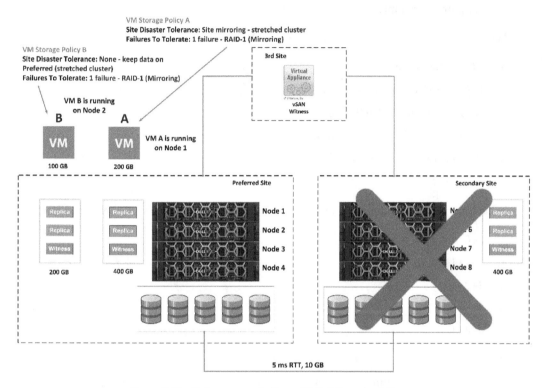

Figure 7.19 – Failure scenario five of VxRail Stretched Cluster

VM A and **VM B** keep running in this scenario, but the vSAN object status of **VM A** will be degraded because **VM Storage Policy A** is configured to **Site mirroring - stretched cluster**. The vSAN object status of **VM B** is healthy because **VM Storage Policy B** is configured to **None - keep data on Preferred (stretched cluster)**.

With the preceding scenario, you now understand the expected results of VxRail Stretched Cluster.

Summary

In this chapter, you saw an overview and learned about the design of VxRail Stretched Cluster, including the network, hardware, and software requirements, and some failure scenarios. When you plan to design an active-active data center or site failure solution, VxRail Stretched Cluster is a good option.

In the next chapter, you will learn about the design of VxRail with VMware SRM and the best practices for implementing this solution.

Questions

The following are a short list of review questions to help reinforce your learning and help you identify areas which require some improvement.

1. Which network bandwidth can be supported on VxRail Stretched Cluster?

    A. 1 GB network bandwidth

    B. 1 GB and 10 GB network bandwidth

    C. 10 GB network bandwidth only

    D. 10 GB and 25 GB network bandwidth

    E. 25 GB network bandwidth only

    F. All of these

2. What is the minimum number of nodes configuration supported with VxRail Stretched Cluster?

    A. 1+0+1

    B. 1+1+1

    C. 2+2+1

    D. 3+3+1

    E. 4+4+1

    F. 5+5+1

3. What is the maximum number of nodes configuration supported with VxRail Stretched Cluster?

    A. 5+5+1

    B. 8+7+1

    C. 10+10+1

    D. 12+13+1

    E. 15+15+1

    F. 20+20+1

4. Which VxRail software releases can support VxRail vCenter Server on VxRail Stretched Cluster?

    A. VxRail 4.7.200

    B. VxRail 4.7.300

    C. VxRail 4.7.410

    D. VxRail 7.0.240

E. VxRail 7.0.300

F. All of these

5. What is the maximum supported RTT between VxRail Stretched Cluster and the vSAN witness?

A. 100 milliseconds

B. 200 milliseconds

C. 300 milliseconds

D. 400 milliseconds

E. 500 milliseconds

F. All of these

6. How many FDs need to be created in VxRail Stretched Cluster?

A. One FD

B. Two FDs

C. Three FDs

D. Four FDs

E. Five FDs

F. Six FDs

7. Which network setting is used to separate the vSAN network traffic and witness network traffic?

A. NIOC

B. WTS

C. vSphere Standard Switch

D. vSphere Distributed Switch

E. DRS

F. None of these

8. Which setting is supported with site mirroring in the VM storage policy?

A. None - keep data on Preferred

B. None - keep data on Secondary

C. Site mirroring - stretched cluster

D. None - standard cluster

E. None - stretched cluster

F. None of these

9. Which setting can support three vSAN nodes' failures that can be selected in the VM storage policy? Assume the three nodes are in a single site.

    A. RAID-1 (Mirroring)

    B. RAID-0

    C. RAID-5 (Erasure Coding)

    D. RAID-6 (Erasure Coding)

    E. RAID-10 (Mirroring)

    F. RAID-5/6 (Erasure Coding)

10. What status will the VMs trigger if the vSAN communication is disconnected between the preferred and secondary sites?

    A. The VMs keep running.

    B. All of the VMs shut down.

    C. The VMs on the preferred FD will shut down and power on the secondary FD.

    D. The VMs on the secondary FD will shut down and power on the preferred FD.

    E. None of these.

11. What status will the VMs trigger if the vSAN communication is disconnected between data nodes (preferred and secondary sites) and the witness in VxRail Stretched Cluster?

    A. The VMs keep running.

    B. All of the VMs shut down.

    C. The VMs on the preferred FD will shut down and power on the secondary FD.

    D. The VMs on the secondary FD will shut down and power on the preferred FD.

    E. None of these.

12. Which vSAN license edition/s supports VxRail Stretched Cluster?

    A. vSAN Standard edition

    B. vSAN Advanced edition

    C. vSAN Enterprise edition only

    D. vSAN Enterprise and Enterprise Plus editions

    E. vSAN Enterprise Plus only

    F. All of these

8
Design of VxRail with SRM

In the previous chapter, you had an overview of the design of a VxRail stretched cluster, including network, hardware, and software requirements, and some failure scenarios. When you plan to design an active-active data center solution or site failure solution, a VxRail stretched cluster is a good option.

VxRail also supports **disaster recovery** (**DR**) and data replication solutions, including **VMware Replication** (**VR**), VMware **Site Recovery Manager** (**SRM**), and Dell EMC **RecoverPoint for Virtual Machines** (**RP4VMs**). Compared to a VxRail stretched cluster, VxRail VMware SRM is an active-passive solution. This chapter will discuss the integration of VxRail with VMware SRM and the best practices relating to this solution. In *Chapter 9*, *Design of RecoverPoint for Virtual Machines on VxRail*, we will discuss the design of Dell EMC RP4VMs.

This chapter covers the following main topics:

- Overview of VMware SRM
- Overview of VxRail with VMware SRM
- The design of VxRail with VMware SRM
- The benefits of VxRail with VMware SRM
- Failover scenarios of VxRail with VMware SRM

Overview of VMware SRM

We will start by going through an overview of VMware SRM. VMware SRM is a DR and data replication solution. It supports two protection methodologies: **VM replication** and **storage replication**. *Figure 8.1* shows a conceptual overview of VMware SRM. There are two sites (**Protected Site** and **Recovery Site**), and each site contains a protected workload and a non-protected workload. In a protected workload, you can define the SRM recovery plan by using **vSphere Replication** and **storage replication**, with both configurations in the recovery site. If the protected site is down, you can execute the SRM recovery plan to recover the VMs in the recovery site:

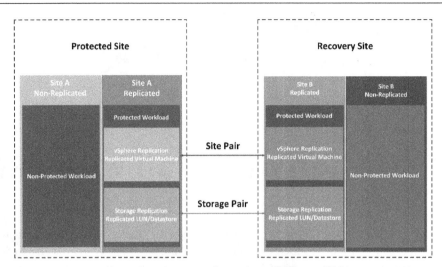

Figure 8.1 – Conceptual overview of VMware SRM

With the preceding information, you should understand how VMware SRM works. Now, we will discuss the architecture of the VMware SRM solution.

Figure 8.2 shows a logical diagram of VMware SRM:

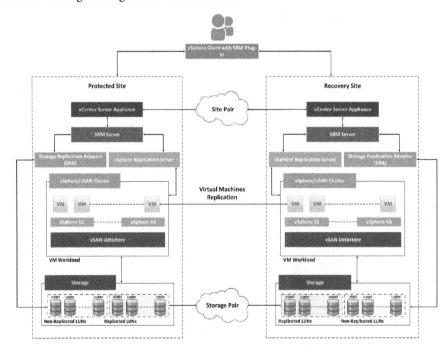

Figure 8.2 – A logical diagram of VMware SRM

There are two sites: **Protected Site** and **Recovery Site**. In each site, there are four main components: **vCenter Server Appliance**, **SRM Server**, **Storage Replication Adapter (SRA)**, and **vSphere Replication Server**. Now, we will discuss the details of each component:

- **vCenter Server Appliance** is used to manage all operational tasks of VMware SRM—for example, creating an SRM recovery plan, executing an SRM failover test, and carrying out the SRM recovery plan.

- **SRM Server** is a virtual appliance that can deliver DR and data replication solutions. SRM can support two protection methods: VM replication and storage-based replication. You can choose one or both methods to define the DR plan based on the requirements of your **Recovery Time Objective (RTO)** and **Recovery Point Objective (RPO)**.

- **Site Recovery Adapter (SRA)** is software that is provided by the storage vendor. It can work with SRM and storage and deliver storage-based replication across the protected and recovery sites.

- **vSphere Replication (VR) Server** is a virtual appliance that delivers VM replication at a local site or both protected and recovery sites. It can support a minimum of 5-minute RPO of VM replication.

When you deploy SRM into your vSphere environment, you can choose from the following deployment types. You must deploy the SRM virtual appliance in a vCenter Server environment by using the **Open Virtualization Format (OVF)** deployment wizard. *Table 8.1* shows the differences between each deployment type:

Deployment Type	Requirements
Light	Two vCPUs, 8 GB memory, one 16 GB VMDK and one 4 GB VMDK, and one 1 GB network adapter. This deployment supports the protection of fewer than 1,000 VMs.
Standard	Four vCPUs, 12 GB memory, one 16 GB VMDK and one 4 GB VMDK, and one 1 GB network adapter. This deployment supports the protection of more than 1,000 VMs.

Table 8.1 – The deployment types of VMware SRM

You need to open the firewall ports for each core component during SRM configuration across two sites. *Figure 8.3* shows the network ports for the VMware SRM appliance, along with the required network ports for each core component:

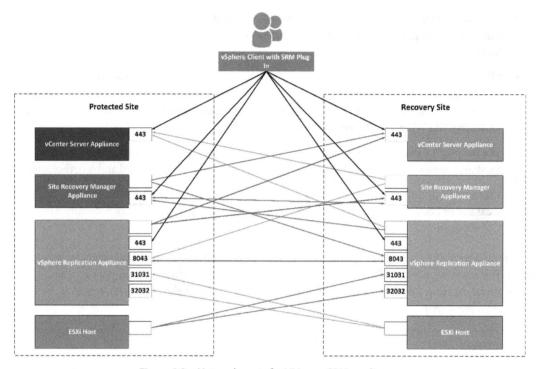

Figure 8.3 – Network ports for VMware SRM appliance

Now, we will list the network port requirement for SRM, VR, vCenter Server, and the ESXi host. *Table 8.2* shows a summary of the SRM port requirements for vCenter Server and the ESXi host:

Port	Protocol	Source	Target	Description
443	HTTPS	SRM	vCenter Server	**Secure Sockets Layer (SSL)** web port
443	HTTPS	SRM at the recovery site	The ESXi hosts at the recovery site	The network traffic from the SRM on the recovery site to the ESXi hosts when you execute the SRM recovery plan or failover test

Table 8.2 – Summary of SRM port requirements for vCenter and the ESXi host

Table 8.3 shows a summary of SRM's network port requirements:

Port	Protocol	Source	Target	Description
443	HTTPS	SRM	vCenter Server	SSL web port for incoming TCP network traffic
443	HTTPS	SRM at the recovery site	The ESXi hosts at the recovery site	The network traffic from SRM on the recovery site to the ESXi hosts when you execute the SRM recovery plan or failover test

Table 8.3 – Summary of SRM's network port requirements

Table 8.4 shows a summary of the SRM and VR network port requirements at the protected and recovery sites:

Port	Protocol	Source	Target	Description
31031	Replication network traffic	ESXi host	VR at the recovery site	The ESXi hosts in the protected site to the VR in the recovery site
32032	TCP	SRM at the protected site	VR at the recovery site	The initial and outgoing replication traffic from the ESXi hosts at the protected site to the ESXi hosts at the recovery site
8043	HTTPS	VR at either the protected site or recovery site	VR at either the protected site or recovery site	The management traffic between VR appliances
8043	HTTPS	SRM	VR at both the protected site and recovery site	The management traffic between the VR appliance and SRM

Table 8.4 – Summary of SRM and VR network port requirements

Table 8.5 shows a summary of VR and additional VR Server network port requirements at the protected and recovery sites:

Port	Protocol	Source	Target	Description
80	HTTP	VR appliance	ESXi host	To make the connection before initiating the VM replication
443	HTTPS	SRM HTML5 interface	VR appliance	The default port is the SRM HTML5 interface
5480	HTTPS	Browser	VR appliance	VR **Virtual Appliance Management Interface (VAMI)** website UI
31031	The initial and ongoing replication traffic	ESXi host at the protected site	VR or external VR at the recovery site	The initial and outgoing replication traffic from the ESXi hosts at the protected site to the ESXi hosts at the recovery site
32032	TCP	ESXi host at the protected site	VR at the recovery site	The initial and outgoing replication traffic from the ESXi hosts at the protected site to the ESXi hosts at the recovery site

Table 8.5 – Summary of VR and additional VR server network port requirements at the protected and recovery sites

When you execute the site pairing for SRM between the protected and recovery sites, port 443 needs to open. *Table 8.6* shows a summary of SRM site pairing between the protected and recovery sites:

Port	Protocol	Source	Target	Description
443	HTTPS	vCenter Server	SRM appliance	The communication between vCenter Server and the SRM appliance
443	HTTPS	SRM appliance	SRM appliance at the recovery site	The communication between each SRM appliance
443	HTTPS	SRM appliance	vCenter Server	The communication between vCenter Server and the SRM appliance at the protected and recovery sites

Table 8.6 – Summary of SRM site pairing between the protected and recovery sites

In the next section, we will discuss an overview of VM replication with SRM.

VM replication

If you choose VM replication for the DR solution, you need to consider the following:

- VR is a bundled feature that includes the license in the following editions of VMware vSphere:

    - vSphere Essentials Plus

    - vSphere Standard

    - vSphere Enterprise

    - vSphere Enterprise Plus

    - vSphere Desktop

- SRM licenses are required for at least 25 VMs; they support one-way protection from the protected site to the recovery site.

- In SRM 8.4 and above, SRM deployment on windows is no longer available. We need to use appliances.

- You can only deploy one VR appliance into a vCenter Server instance.

- For SRM configurations, one VR appliance for vCenter is supported.

- The VR appliance can support a maximum of 3,000 VM replications.

- You need to deploy one vCenter Server appliance, one SRM appliance, and one VR appliance in the protected and recovery sites.

- The minimum RPO of the protected VM is 5 minutes per vCenter Server instance.

- VR can also be supported with certain vSphere features—that is, vSphere vMotion and vSphere High Availability.

- For the bandwidth requirements of VR, the RPO can affect the data change rate. You need to evaluate how many blocks change in your RPO for the protected VM. To calculate the bandwidth for VR, you can use this link: `https://docs.vmware.com/en/vSphere-Replication/8.5/com.vmware.vsphere.replication-admin.doc/GUID-4A34D0C9-8CC1-46C4-96FF-3BF7583D3C4F.html`.

In *Figure 8.4*, there are two sites: protected and recovery sites. There are three main components per site, **vCenter Server Appliance**, **SRM Server**, and **vSphere Replication Server**. You can create VR protection groups to replicate the VMs from either a protected site or a recovery site. You can execute the SRM recovery plan to recover the VMs into the recovery site if the protected site is faulty:

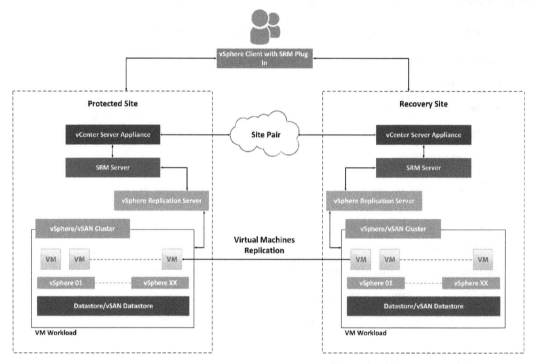

Figure 8.4 – A logical diagram of VMware SRM with VR

> **Important note**
>
> For compatibility matrices of VR 8.5.x, you can check the details at https://docs.vmware.com/en/vSphere-Replication/8.5/rn/vsphere-replication-compat-matrix-8-5.html.

In the next section, we will see an overview of storage-based replication with SRM.

Storage-based replication

When you choose storage-based replication for the DR solution, you need to consider the following:

- SRM licenses are required for at least 25 VMs; they support one-way protection from the protected site to the recovery site.

- The source and target storage systems should be identical or the same model, and storage-based replication is enabled between the protected site and the recovery site.

- Storage-based replication supports Fibre Channel and iSCSI connections between the source and target storage.

- In SRM 8.4 and above, SRM deployment on Windows is no longer available. We need to use appliances.

- You need to deploy one vCenter Server appliance, one SRM appliance, and one VR appliance in the protected and recovery sites.

- For the storage-based replication traffic requirements, you need to consult the relevant storage vendors.

- You need to verify the compatibility with the SRM release version and the storage vendor's SRA. For more details, you can use this link: `https://www.vmware.com/resources/compatibility/search.php?deviceCategory=sra`.

Next, we will discuss how to verify the compatibility of SRM and SRA in the *VMware Compatibility Guide*.

Figure 8.5 shows the *VMware Compatibility Guide*; you select the SRM version from the **Product Release Version** menu, the storage vendor from the **Partner Name** menu, and SRA from the **SRA Name** menu in the **VMware Compatibility Guide** wizard, and then it can list the supported versions of SRM and SRA. In this example, **SRM 8.5/8.4/8.3/8.2** can support **EMC Unity Block SRA 5.0.4.146**:

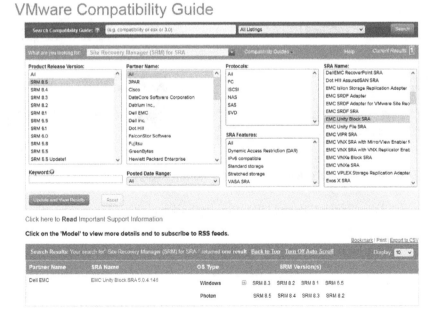

Figure 8.5 – VMware Compatibility Guide

Figure 8.6 shows a logical diagram of VMware SRM with storage-based replication. There are two sites: protected and recovery. There are three main components per site, **vCenter Server Appliance**, **SRM Server**, and **Site Recovery Adapter**. You can use storage-based replication to replicate the data between the protected site and the recovery site, and you must use SRA to integrate the SRM and storage for this data replication. You can execute the SRM recovery plan to recover the VMs into the recovery site if one of the sites is faulty. If the storage supports **Consistency Groups** (**CGs**), SRM is supported with vSphere Storage DRS and vSphere Storage vMotion:

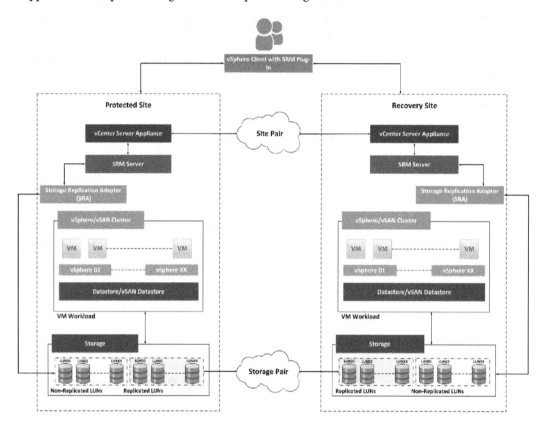

Figure 8.6 – A logical diagram of VMware SRM with storage-based replication

The next section will discuss the VMware SRM licensing package.

SRM licensing

VMware SRM is available in two editions, **Standard** and **Enterprise**. Each SRM license pack includes 25 VMs. Only one edition of SRM can be configured in a vCenter Server instance. VMware SRM Standard edition is used for smaller environments and is limited to 75 protected VMs per site. VMware SRM Enterprise edition is used for the enterprise-level protection of VMs, and there's no limit on the number of protected VMs. *Table 8.7* shows a feature comparison of the SRM Standard and Enterprise editions:

SRM Licensing/Features	Standard Edition	Enterprise Edition
Maximum protected VMs	Supported	Supported
Integration with VMware Cloud products	Supported	Supported
Centralized recovery plans	Supported	Supported
Non-disruptive recovery test	Supported	Supported
Automated DR failover	Supported	Supported
Planned data center migration	Supported	Supported
Automated re-protection and failback	Supported	Supported
Storage-based replication support	Supported	Supported
Automatic protection of VMs	Supported	Supported
VR support	Supported	Supported
Stretched storage support	Not supported	Supported
Orchestrated cross-vCenter vMotion	Not supported	Supported
Virtual volumes support	Not supported	Supported

Table 8.7 – Features comparison of SRM Standard and Enterprise editions

With the preceding information on VMware SRM, you should now have an understanding of VR and storage-based replication with SRM. *Table 8.8* shows a comparison of VR and storage-based replication with SRM:

Features	VM Replication with SRM	Storage-Based Replication with SRM
VMware SRM license	At least 25 VMs	At least 25 VMs
Hypervisor-based replication	Yes	No
RPO	5 minutes (minimum)	Depends on the storage model
Multi-VM CG	Not supported	Supported
vMotion and Storage vMotion	Supported	Not supported
Built-in WAN compression	Supported	Depends on the storage model
Automatically protect new VMs	Supported	Supported
Failback with reverse protection	Supported	Supported
Non-disruptive failover test	Supported	Supported
Automatic re-IP of VMs	Supported	Supported
Recovery failover reports	Supported	Supported

Table 8.8 – Comparison of VR and storage-based replication with SRM

With the preceding information, you should now have a good understanding of SRM. The next section will provide an overview of VxRail with VMware SRM.

Overview of VxRail with VMware SRM

When enabling the DR features on the VxRail cluster, you can choose from VR, SRM, or Dell EMC RP4VMs, or from third-party data protection solutions. Since VxRail is developed by Dell and VMware, most VMware solutions are fully integrated with VxRail. *Figure 8.7* shows a logical diagram of VxRail with VMware SRM. There are two sites: protected and recovery. This solution includes four main components per site—that is, **VxRail Cluster**, **vCenter Server Appliance**, **SRM Server**, and **Site Recovery Adapter**:

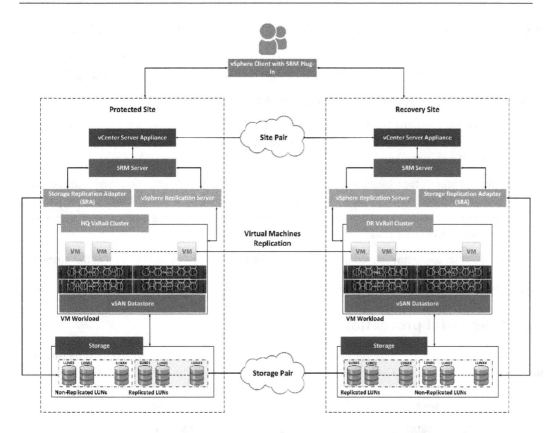

Figure 8.7 – A logical diagram of VxRail with VMware SRM

Based on your RTO and RTP requirements, you can choose VM replication with SRM or storage-based replication with SRM for the DR solution. **HQ VxRail Cluster** and **DR VxRail Cluster** can be the same or different series of VxRail systems. For example, an HQ VxRail cluster is created with four VxRail P670 nodes, and a DR VxRail cluster is created with four VxRail E660 nodes. vCenter Server can be an embedded vCenter Server or a customer-supplied vCenter Server, and the embedded vCenter Server is recommended because the vCenter Server Standard license is bundled on the VxRail cluster. If you choose the customer-supplied vCenter Server, it requires the optional vCenter Server Standard license for each vCenter Server. VxRail supports access to external storage with Fibre Channel or iSCSI. If you choose storage-based replication with SRM, please make sure the storage replication and snapshot features are enabled on the source and target storage; otherwise, the operation task of the SRM recovery plan will fail.

If you choose VxRail with VMware SRM for the DR solution, you need to consider the following:

- This DR solution supports VxRail All-Flash and Hybrid nodes. For example, the source VxRail cluster can be the All-Flash nodes and the target VxRail cluster can be the Hybrid nodes.

- This DR solution supports both VM replication and storage-based replication.

- VMware SRM includes two types of licensing packages: the Standard edition and the Enterprise edition. If the number of protected VMs is less than 75 per site, you can choose the Standard edition. If it is more than 75 VMs, you need to choose the Enterprise edition.

- If the VM and storage-based replication provides uni-directional protection (one-way) across the protected and recovery sites, you need one SRM license pack (25 VM licenses per pack). If it provides bi-directional protection (two-way) across the protected and recovery sites, you need two SRM license packs.

- For the other requirements, you can refer to the *Overview of VMware SRM* section of this chapter.

Now, we will discuss the difference between uni-directional protection and bi-directional protection.

Uni-directional protection

Uni-directional protection is used in failover use cases only. *Figure 8.8* shows the architecture of uni-directional protection in SRM; SRM is only configured to fail over the VMs from **Site A** to **Site B**:

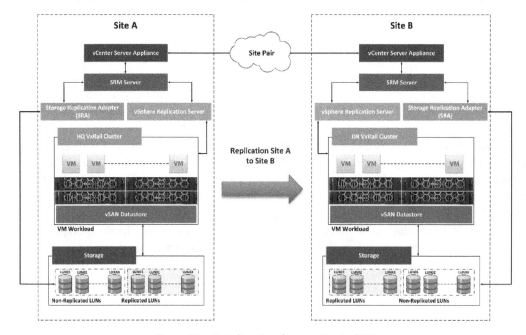

Figure 8.8 – Uni-directional protection in SRM

Table 8.9 shows the required VMware licenses for uni-directional protection:

License	Site A	Site B
vCenter Server	vCenter Server Standard edition	vCenter Server Standard edition
vSphere	vSphere Standard/Enterprise Plus edition	vSphere Standard/Enterprise Plus edition
VR	It includes vSphere Essentials Plus and higher editions	It includes vSphere Essentials Plus and higher editions
SRM	SRM Standard/Enterprise edition	No SRM license required

Table 8.9 – The required VMware licenses for uni-directional protection

Bi-directional protection

Bi-directional protection is used in failover from **Site A** to **Site B** and **Site B** to **Site A** at the same time. *Figure 8.8* shows the architecture of bi-directional protection in SRM; the SRM license must add to **Site A** and **Site B** if the replication is enabled in two ways:

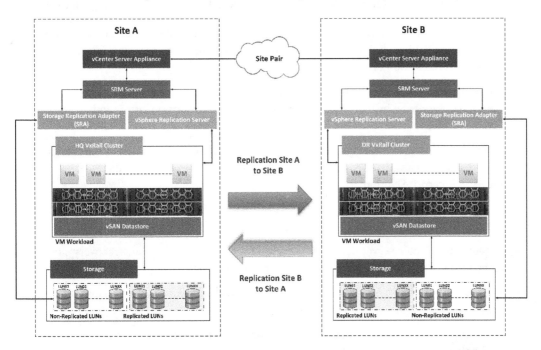

Figure 8.9 – Bi-directional protection in SRM

Table 8.10 shows the required VMware licenses for bi-directional protection:

License	Site A	Site B
vCenter Server	vCenter Server Standard edition	vCenter Server Standard edition
vSphere	vSphere Standard/Enterprise Plus edition	vSphere Standard/Enterprise Plus edition
VR	It includes vSphere Essentials Plus and higher editions	It includes vSphere Essentials Plus and higher editions
SRM	SRM Standard/Enterprise edition	SRM Standard/Enterprise edition

Table 8.10 – The required VMware licenses for bi-directional protection

The next section will discuss the design of VxRail with VMware SRM.

The design of VxRail with VMware SRM

In this section, we will discuss the planning and design of VxRail with VMware SRM. When you plan a DR solution for VxRail with VMware SRM, you need to consider and collect the following information:

- What are your RPO and RTO requirements for DR? These are both important factors you need to keep in mind while designing and executing a DR plan. Based on these two factors, you can choose VM replication or storage-based replication.

- The VxRail All-Flash and Hybrid nodes are both supported in this solution. For example, the source cluster can be the VxRail All-Flash cluster and the target cluster can be the VxRail Hybrid cluster.

- How many protected VMs are required for recovery at the recovery site? If the number of protected VMs is less than 75, you can choose the SRM Standard edition; otherwise, you need the SRM Enterprise edition.

- What is the total usable storage of protected VMs? You need to prepare the same usable storage capacity in the VxRail cluster in the recovery site.

- What is the vCenter Server deployment option for the protected site and recovery site—an embedded vCenter Server or customer-supplied vCenter Server?

- If you choose storage-based replication, make sure the data replication and snapshot features are enabled on the source and target storage.

- If you choose the VM replication, it supports deploying multiple VR instances to meet your load-balancing needs at each site.

- What is the replication network traffic between the protected and recovery sites?

- How many VMs and protection groups are associated with an SRM recovery plan?

- What is the running edition of the VxRail software release in your existing environment?

- It is highly recommended that VMware Tools is installed in all protected VMs in the protected site.

- You need to prepare a dedicated datastore for placeholder VMs at the protected and recovery sites. When you create a protection group for the VMs at the protected site, SRM creates placeholder VMs at the recovery site for each protected VM.

- Each VM requires a VM swap file. By default, VM files and VM swap files are stored in the same datastore. To prevent SRM from replicating the swap files in the recovery sites, you can create an unreplicated datastore for swap files. This can help avoid wasting network bandwidth during replication between the protected and recovery sites.

You need to collect the preceding information when planning to use VxRail with VMware SRM. Next, let's look at the benefits of VxRail with VMware SRM.

The benefits of VxRail with VMware SRM

VxRail with VMware SRM can deliver the following key benefits and capabilities:

- It can perform non-interruptive automated failover and recovery tests in an isolated network during business hours.

- It can perform a DR failover or a planned migration and recover a VM to the original site with a *one-click* operation. It can also reduce recovery time using automated orchestration workflows.

- An SRM recovery plan supports recovering VMs with a different IP address; it also supports extending the same subnets via integration with VMware **Network Virtualization** (**NSX**) at the recovery site.

- You can manage and create an SRM recovery plan for a set of VMs via a central management HTML5 graphic UI.

- It can deliver a failover and recovery test report for the protected VMs.

- SRM can integrate with VR and deliver data recovery with a 5-minute RPO.

- VR supports VM replication with network compression across the protected and recovery sites, and it can minimize the network bandwidth consumption between the two sites.

- SRM can support a range of storage-based replication solutions.

- VxRail is fully supported by VMware products, and it natively uses the benefits of VMware Cloud Foundation, VMware NSX, and VMware vSAN.

- SRM and VR can extend to a **DR as a Service** (**DRaaS**) that offers VMware Cloud on **Amazon Web Services** (**AWS**).

- VxRail also supports other DR and data replication solutions; SRM and VR are examples. If your RPO and RTO require an active-active architecture, you can choose the VxRail stretched cluster; refer to *Chapter 7, Design of Stretched Cluster on VxRail*. If you need the **Continuous Data Protection** (**CDP**) solution, refer to *Chapter 9, Design of RecoverPoint for Virtual Machines on VxRail*.

> **Important note**
>
> If you enable VMware Cloud Foundation and VMware NSX on VxRail with VMware SRM, it requires optional licenses for VMware Cloud Foundation and VMware NSX.

The next section will discuss different failover scenarios of VxRail with VMware SRM.

Failover scenarios of VxRail with VMware SRM

This solution is not only used for DR features; it can also be used for data migration. This section will discuss some failover scenarios of VxRail with SRM.

Failure scenario 1

Figure 8.10 shows two different environments: one is a **vSphere cluster** and the other is a **VxRail cluster**. Each site includes the following components:

- **Site A**: vCenter Server Appliance, SRM Server, vSphere Replication Server, vSphere Cluster with four nodes, and **Storage**.

- **Site B**: vCenter Server Appliance, SRM Server, vSphere Replication Server, and **VxRail Cluster** with four nodes.

- **Replication configuration**: The VMs are replicated to **Site B** from **Site A** with VR:

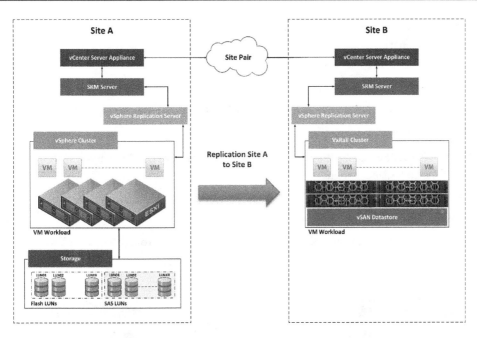

Figure 8.10 – Scenario 1 of VxRail with VMware SRM

In this scenario, we plan to migrate the VMs into **VxRail Cluster** from **vSphere Cluster**. The SRM recovery plan supports two recovery types, **Planned migration** and **Disaster recovery**, as shown in the following screenshot. We can choose the **Planned migration** option to migrate the VMs into the VxRail cluster from the vSphere cluster:

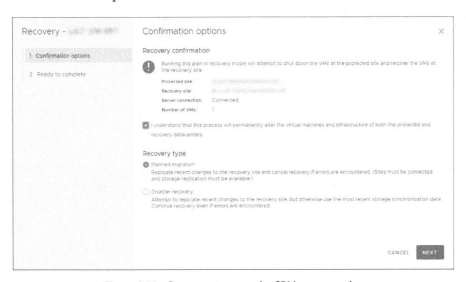

Figure 8.11 – Recovery type on the SRM recovery plan

Using SRM's planned migration, we can easily migrate the VMs into the VxRail cluster. Now, we will discuss the next scenario.

Failure scenario 2

Figure 8.12 shows two VxRail cluster environments, **Site A** and **Site B**. Each site includes the following components:

- **Site A**: **vCenter Server Appliance**, **SRM Server**, **vSphere Replication Server**, **VxRail Cluster** with four nodes, and storage.

- **Site B**: **vCenter Server Appliance**, **SRM Server**, **vSphere Replication Server**, and **VxRail Cluster** with four nodes.

- **Replication configuration**: The VMs are replicated to **Site B** from **Site A** with VR:

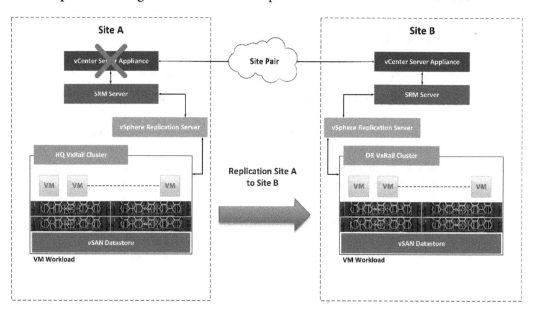

Figure 8.12 – Scenario 2 of VxRail with SRM and VR

If the vCenter Server appliance is faulty at **Site A**, it does not affect the replication of the VMs between **Site A** and **Site B**. You can still manage the VM replication through vCenter Server in **Site B**.

Failure scenario 3

Figure 8.13 shows two VxRail cluster environments, **Site A** and **Site B**. Each site includes the following components:

- **Site A**: **vCenter Server Appliance**, **SRM Server**, **vSphere Replication Server**, **VxRail Cluster** with four nodes, and storage.

- **Site B**: **vCenter Server Appliance**, **SRM Server**, **vSphere Replication Server**, and **VxRail Cluster** with four nodes.

- **Replication configuration**: The VMs are replicated to **Site B** from **Site A** with VR:

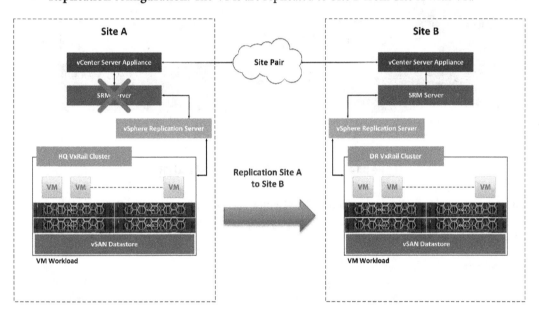

Figure 8.13 – Scenario 3 of VxRail with SRM and VR

If the SRM server is faulty at **Site A**, it does not affect the replication of VMs between **Site A** and **Site B**. You can still manage the VM replication through vCenter Server in **Site A** or **Site B**.

Failure scenario 4

Figure 8.14 shows two VxRail cluster environments, **Site A** and **Site B**. Each site includes the following components:

- **Site A**: **vCenter Server Appliance**, **SRM Server**, **vSphere Replication Server**, **VxRail Cluster** with four nodes, and storage.

- **Site B: vCenter Server Appliance, SRM Server, vSphere Replication Server**, and **VxRail Cluster** with four nodes.

- **Replication configuration**: The VMs are replicated to **Site B** from **Site A** with VR:

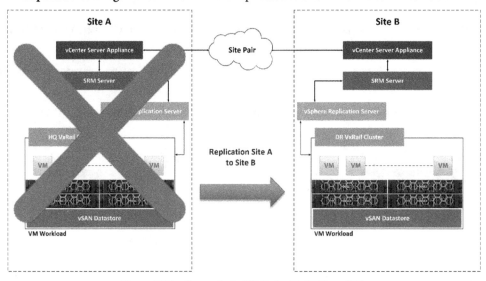

Figure 8.14 – Scenario 4 of VxRail with SRM and VR

If **Site A** is completely faulty, we can access the vCenter Server appliance at **Site B**, then execute the SRM recovery plan and select the **Discover recovery** option shown here to fail over the VMs into the VxRail cluster from **Site A** to **Site B**:

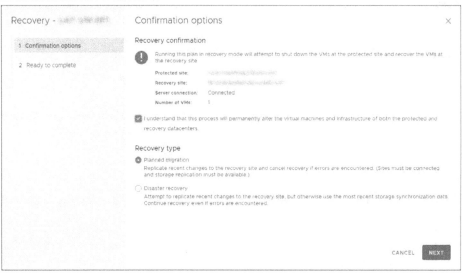

Figure 8.15 – Recovery type on the SRM recovery plan

In this scenario, we can still manage and execute the SRM recovery plan through the vCenter Server appliance if one of the sites is faulty.

Failure scenario 5

Figure 8.16 shows two VxRail cluster environments, **Site A** and **Site B**. Each site includes the following components:

- **Site A**: **vCenter Server Appliance**, **SRM Server**, **vSphere Replication Server**, **VxRail Cluster** with four nodes, and **Storage**.

- **Site B**: **vCenter Server Appliance**, **SRM Server**, **vSphere Replication Server**, and **VxRail Cluster** with four nodes, and **Storage**.

- **Replication configuration**: The VMs are replicated to **Site B** from **Site A** with VR.

- Storage-based replication is enabled between **Site A** and **Site B**:

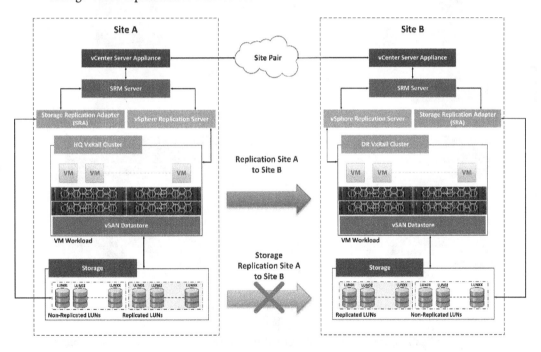

Figure 8.16 – Scenario 5 of VxRail with SRM and storage-based replication

If the storage-based replication links are disconnected between **Site A** and **Site B**, we can execute the SRM recovery plan with already replicated data and recover the VMs into **Site B** based on the latest RPO on storage-based replication, but it won't replicate any data after disconnecting storage replication links.

Failure scenario 6

Figure 8.17 shows two VxRail cluster environments, **Site A** and **Site B**. Each site includes the following components:

- **Site A: vCenter Server Appliance, SRM Server, vSphere Replication Server**, **VxRail Cluster** with four nodes, and **Storage**.

- **Site B: vCenter Server Appliance, SRM Server, vSphere Replication Server**, and **VxRail Cluster** with four nodes, and **Storage**.

- **Replication configuration**: The VMs are replicated to **Site B** from **Site A** with VR.

- Storage-based replication is enabled between **Site A** and **Site B**:

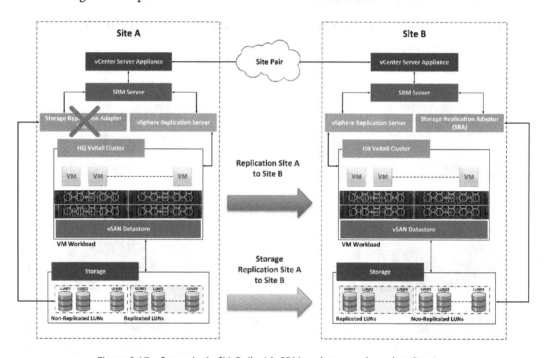

Figure 8.17 – Scenario 6 of VxRail with SRM and storage-based replication

If SRA fails after replication, we can run SRM for the previously replicated data. If SRA fails before initial replication then, we cannot run SRM plans because we won't have any data in **Site B logical unit numbers (LUNs)**.

From the preceding scenarios, you have learned about the failover behavior and use cases for VxRail with VMware SRM. *Table 8.11* shows the expected result of each failover scenario on SRM:

Case #	vCenter Server	SRM/SRA	VR	Storage Replication	Recovery Type	Failover
1	Healthy	Healthy	Healthy	N/A	Planned migration	Successful
2	Faulty (Site A)	Healthy	Healthy	N/A	Discover recovery	Successful
3	Healthy	Faulted (Site A)	Healthy	N/A	Discover recovery	Not successful
4	Faulty (Site A)	Faulted (Site A)	Faulted (Site A)	N/A	Discover recovery	Successful
5	Healthy	Healthy	Healthy	Connection is disconnected	Discover recovery	Successful with warnings
6	Healthy	SRA faulted (Site A)	Healthy	Healthy	Discover recovery	Not successful

Table 8.11 – The expected result of each failover scenario on SRM

Summary

In this chapter, we gave you an overview of VxRail with VMware SRM, including the network, hardware, and software requirements, and some failure scenarios. You can now choose the appropriate VM protection option based on your RPO and RTO requirements. SRM can be used in two scenarios—that is, data migration and DR. If you plan to migrate VMs into the VxRail cluster, SRM is a good option.

In the next chapter, you will learn about the design of RP4VMs on VxRail and the best practices of this solution.

Questions

The following are a short list of review questions to help reinforce your learning and help you identify areas which require some improvement.

1. How many vCenter Server appliances are required on the *VxRail with VMware SRM* solution?

    A. One

    B. Two

    C. Three

    D. Four

    E. Five

    F. None of these

2. Which two protection methods are supported by SRM?

    A. Uni-directional

    B. One-way

    C. Bi-directional

    D. Two-way

    E. Three-way

    F. Mirror

3. Which VMware licenses are bundled on the VxRail cluster?

    A. vCenter Server Standard edition

    B. VMware SRM

    C. VMware vSAN

    D. VMware SRA

    E. vSphere Replication

    F. All of the these

4. How many protected VMs (maximum) can be supported with the SRM Standard edition?

    A. 25

    B. 50

    C. 75

    D. 100

    E. 125

    F. Unlimited

5. How many protected VMs (maximum) can be supported with the SRM Enterprise edition?

    A. 25

    B. 50

    C. 75

    D. 100

    E. 125

    F. Unlimited

6. Which features are not supported with the SRM Standard edition?

    A. Stretched storage

    B. Storage-based replication

    C. Orchestrated Cross-vCenter vMotion

    D. Virtual volumes

    E. Automatic protection of VMs

    F. vSphere Replication

7. What is the minimum RPO that can be supported with vSphere Replication?

    A. 1 minute

    B. 5 minutes

    C. 10 minutes

    D. 15 minutes

    E. 20 minutes

    F. 25 minutes

8. How many protected VMs are included in an SRM license pack?

    A. 10

    B. 15

    C. 25

    D. 50

    E. 75

    F. 100

9. If you plan to enable storage-based replication on SRM, what are the requirements for the source and target storage?

    A. Enable the storage replication feature for the source and target storage

    B. Enable the storage snapshot feature for the source and target storage

    C. Enable the storage tiers feature for the source and target storage

    D. Enable the storage replication and snapshot features for the source and target storage

    E. Enable the storage replication and tiers features for the source and target storage

    F. All of these

10. If you plan to migrate VMs into a VxRail cluster from a vSphere cluster using SRM, which recovery option is the best?

    A.   vSphere Storage vMotion

    B.   vSphere Replication

    C.   Planned migration

    D.   Failover

    E.   DR

    F.   Data replication

11. In *Figure 8.18*, VM replication is enabled between **Site A** and **Site B**. If the vCenter Server appliance is faulty at **Site A**, does it impact the SRM recovery plan?

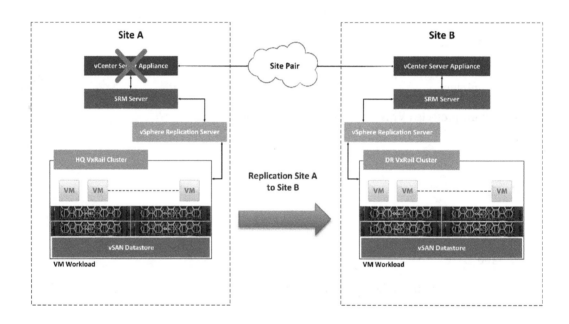

Figure 8.18 – VxRail with VMware SRM

    A.   Yes. We cannot access the vCenter Server appliance.

    B.   No. We can access the vCenter Server appliance at **Site B** and execute the SRM recovery plan.

12. Which of the following SRM configurations shows bi-directional protection?

A.

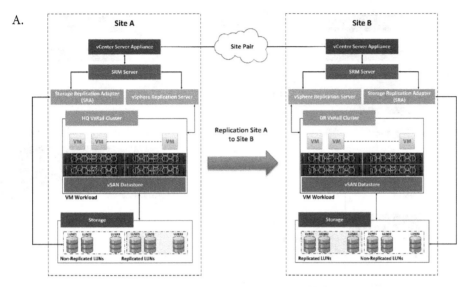

Figure 8.19 – VxRail with VMware SRM

B.

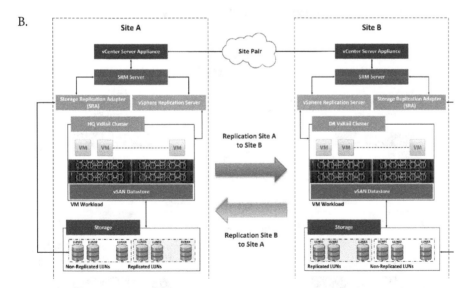

Figure 8.20 – VxRail with VMware SRM

C.

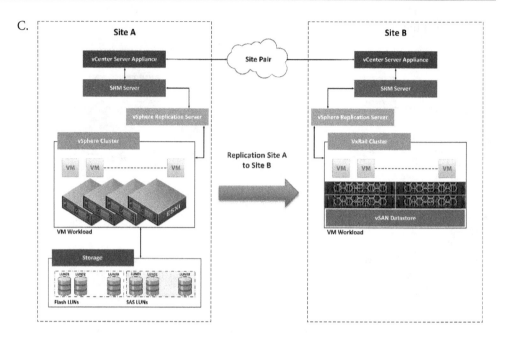

Figure 8.21 – VxRail with VMware SRM

D.

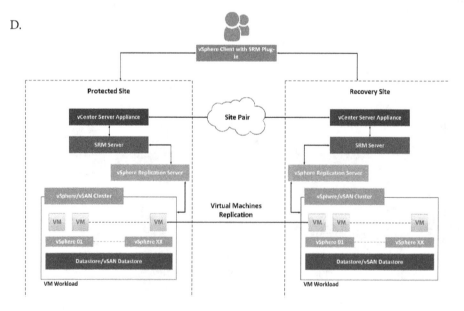

Figure 8.22 – VMware SRM with VR

E.

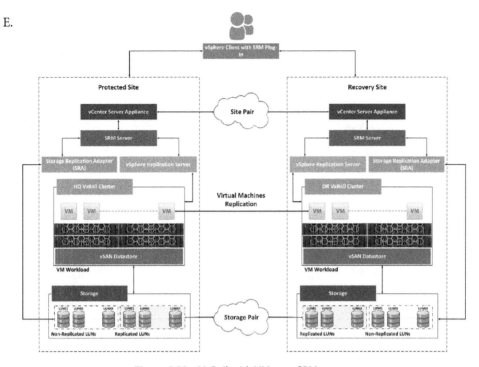

Figure 8.23 – VxRail with VMware SRM

9

Design of RecoverPoint for Virtual Machines on VxRail

In the previous chapter, we covered an overview of VxRail with VMware **Site Recovery Manager** (**SRM**), including the network, hardware, and software requirements, and some failure scenarios. You can choose a **Virtual Machine** (**VM**) protection option based on your **Recovery Time Objective** (**RTO**) and **Recovery Point Objective** (**RPO**) requirements. SRM can be used in two scenarios: data migration and **Disaster Recovery** (**DR**).

In previous chapters, you saw an overview of a vSAN two-node cluster, a VxRail stretched cluster, and VxRail with VMware SRM, and these solutions can provide DR based on your RTO and RPO requirements. **RecoverPoint for Virtual Machines** (**RP4VM**) is a software product powered by Dell; it can deliver **Continuous Data Protection** (**CDP**) for DR. It supports the protection of VMs in the vSphere environment and delivers local and remote replication capabilities at the VM level. Each model of VxRail node comes with limited RP4VM licenses. In this chapter, we will cover an overview of RP4VM, as well as looking at the planning and design of RP4VM on VxRail.

This chapter covers the following main topics:

- Overview of RP4VM
- Overview of RP4VM on VxRail
- The benefits of RP4VM
- The design of RP4VM on VxRail
- Failover scenarios of RP4VM on VxRail

Overview of RP4VM

Before we discuss the design of RP4VM, we will provide an overview of it. *Figure 9.1* shows the RP4VM system architecture with the vSphere HTML5 plugin:

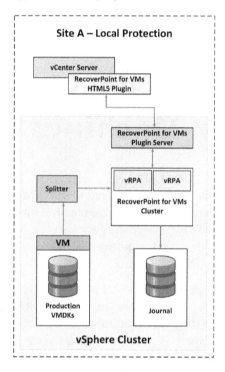

Figure 9.1 – RP4VM system architecture with the vSphere HTML5 plugin

RP4VM includes the following components:

- **VMware vCenter Server**: This central management dashboard provides all operation tasks for RP4VM.

- **RecoverPoint for VMs cluster**: Each RP4VM cluster consists of a minimum of one **Virtual RecoverPoint Appliance (vRPA)** and a maximum of eight vRPAs. This cluster is used to manage the data replication tasks.

- **Splitter**: The RP4VM splitter is installed on each VMware ESXi host. It splits the write tasks coming from the ESXi host and sends them to the vRPA and the VM's **VM disk file (VMDK)**. The vRPA cluster handles all network traffic to journals and replicas.

- **RecoverPoint for VMs HTML5 plugin**: This is the user interface to manage the RP4VM system. The RP4VM 5.3 plugin is the latest version and comes in two versions: the RP4VM plugin and the RP4VM HTML5 plugin. The RP4VM plugin is used for vSphere 6.7 and earlier; it communicates directly with vRPA clusters. The RP4VM HTML5 plugin is available in vSphere 6.7 Update 1 or above; it requires that the vRPA cluster communicates with the HTML5 plugin server.

- **RecoverPoint for VMs HTML5 plugin server**: This server is a virtual appliance and deploys from an **Open Virtualization Application/Appliance (OVA)** through vCenter Server. It manages RP4VM 5.3 or above on vCenter Server. The RP4VM HTML5 plugin server communicates with vCenter Server and the vRPA cluster.

- **Journal**: The journal volume determines how many **point-in-time (PIT)** images of VMs can be stored.

> **Important note**
>
> RP4VM supports vCenter Server in Embedded Linked Mode and supports two configurations. One RP4VM HTML5 plugin server supports one vCenter Server instance where vRPA clusters are registered and one RP4VM HTML5 plugin server to manage all vRPA clusters registered across the different vCenter Server appliances.

RP4VM supports **local protection** and **remote protection** across two sites. *Figure 9.2* shows the RP4VM system architecture with the vSphere HTML5 plugin across two sites:

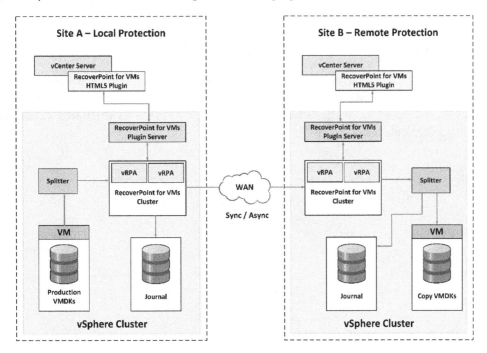

Figure 9.2 – RP4VM system architecture with the vSphere HTML5 plugin across two sites

When we choose RP4VM as a DR solution, we need to consider the following factors:

- The number of vRPAs and the deployment type.

- The number of VMware ESXi hosts.

- The replication type of the vRPA cluster—that is, synchronous replication or asynchronous replication.

- The RTO and RPO requirements for DR.

- What is the running version of vSphere ESXi and vCenter Server?

- How many **Consistency Groups (CGs)** are required?

- The number of network adapters on each vRPA.

- The maximum latency for a remote copy is 10 ms for synchronous replication and 200 ms for asynchronous replication between two sites.

- How many copies are required for the initial configuration?

- The configuration of vRPA includes three configurations: bronze, silver, and gold.

- The size of journal volumes and how many PIT images are stored.

The section will discuss the design of vRPA adapters on a vRPA.

vRPA adapters

Each vRPA contains three virtual interfaces—that is, WAN, LAN, and data. When you deploy a vRPA, you can choose one to four virtual interfaces. This network design can affect the overall network performance, and it cannot be changed after deployment. *Figure 9.3* shows the vRPA network adapter options during the deployment of a vRPA:

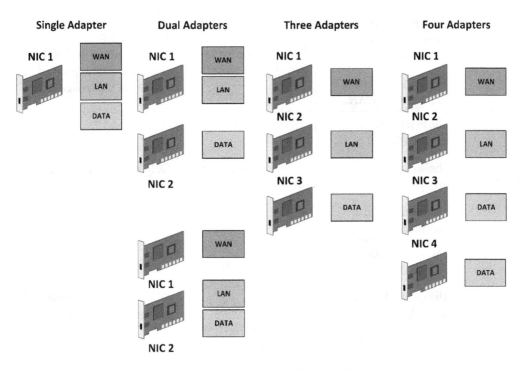

Figure 9.3 – vRPA network adapter options

There are four network adapter options. A single network adapter is used for development or **proof-of-concept** scenarios. You need to separate LAN and WAN interfaces if remote replication is required. The LAN and data interfaces must be on separate subnets.

Table 9.1 shows a comparison of the vRPA adapter options:

Network Adapters	Roles of vRPA Interfaces	Advantages	Disadvantages
One	A single interface for WAN, LAN, and data.	There are fewer IP addresses for configuration.	It does not provide a **High-Availability (HA)** feature; this network design is not used for production environments.
Two	The first interface is a combination of WAN and LAN. The second interface is only used for data.	It can provide better performance and HA. This is the default setting.	N/A.

Two	The first interface is a combination of LAN and data. The second interface is only used for WAN.	It can provide better performance and HA.	**Dynamic Host Configuration Protocol (DHCP)** is not supported on the LAN interface.
Three	The first interface uses WAN. The second interface uses LAN. The third interface uses data.	It can provide improved performance and HA.	DHCP is not supported on the LAN interface.
Four	The first interface uses WAN. The second interface uses LAN. The third and fourth interfaces use data.	It can provide the best performance and HA. The two dedicated network interfaces are used for data.	DHCP is not supported on the LAN interface.

Table 9.1 – A comparison of each vRPA adapter option

vRPA performance profiles

When you deploy a new vRPA into your vSphere environment, you need to select one of the following predefined configurations. These vRPA performance profiles depend on **Input/Output Operations per Second (IOPS)**, the throughput of protected VMs, and the total number of VMs protected by the vRPA cluster:

- **Bronze profile**: Two vCPUs and 8 GB of memory
- **Silver profile**: Four vCPUs and 8 GB of memory
- **Gold profile**: Eight vCPUs and 8 GB of memory

The bronze profile is used for low-throughput scenarios.

The silver profile is used for most situations.

The gold profile is used for high throughput and large environments. Synchronous replication, compression, and deduplication are required.

> **Important note**
> vRPA performance profiles can be changed after deployment by editing the vRPA settings; you can manually allocate more vCPUs and memory to deployed vRPAs.

Table 9.2 shows each vRPA performance profile:

Performance Profile	Supported Features and Maximums
Bronze configuration	It supports up to 128 CGs per vRPA It supports 512 VMs per vRPA It has a total VM writes throughput of up to 70 MB/sec It supports asynchronous replication only
Silver configuration	It supports up to 128 CGs per vRPA It has a total VM writes throughput of up to 200 MB/sec It supports asynchronous replication only
Gold configuration	It supports up to 256 CGs per vRPA It has a total VM writes throughput of up to 350 MB/sec It supports both synchronous and asynchronous replication It supports compression and deduplication

Table 9.2 – A summary of each vRPA performance profile

CGs

CGs are a collection of protected VMs in a group to determine which VMs can be recovered to the same PIT or maintain the correct protection level. There are two replication modes in the CG settings: **Synchronous** and **Asynchronous**. If you choose **Synchronous** in the CG settings, the RPO will be set to 0, but this setting may impact the performance of the protected VMs. If you choose **Asynchronous** in the CG settings, you can define the RPO settings in seconds, and this setting does not affect the performance of the protected VMs. Different factors determine the RPO settings—for example, journal volume performance, WAN link latency, and copy performance. *Figure 9.4* shows the CG settings in RP4VM. RP4VM supports compressing the snapshots in the journal volumes, and this compression impacts the CPU resources of the target vRPA of the CG when you enable journal compression:

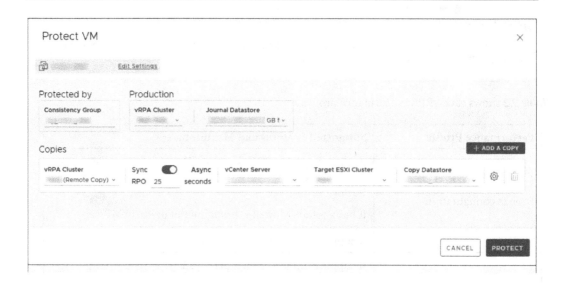

Figure 9.4 – CG settings

Journal volumes

A journal volume is used to store all PITs of each VM image. The RP4VM cluster will automatically build a journal volume on the datastore after the vRPA cluster deployment. The performance of the journal volume is the same as the datastore used for the production VM because all write I/Os are stored in the journal volume. *Figure 9.5* shows the architecture of a journal volume on a vRPA cluster:

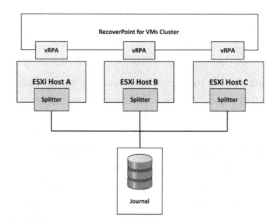

Figure 9.5 – The vRPA cluster for a journal volume

With the preceding information, you have now had an overview of RP4VM and the design of core components. The next section will discuss integrating RP4VM into a VxRail cluster.

Overview of RP4VM on VxRail

The first chapter of this book mentioned how each model of the VxRail bundles some licenses of RP4VM. You can enable the RP4VM feature at any time. If you need CDP for the VMs on a VxRail cluster, you can choose RP4VM. In this solution, you can enable the quick data recovery of VMs at any PIT on a VxRail cluster. When you enable RP4VM on a VxRail cluster, it provides the following key features:

- It supports protecting the VM with recovery at the VM level.

- It supports replicating the VM with VMDK and **raw device mapping** (**RDM**) locally and remotely.

- It uses automation and orchestration to enable test, failover, and failback operations of the VM to any PIT.

- It supports optimizing the WAN bandwidth with data compression and deduplication.

- It supports replicating policies synchronously, asynchronously, or dynamically.

- Using the familiar vSphere HTML5 user interface, you can manage data protection sessions.

- The CGs support crash-consistent and application-consistent recovery of VMs. The RecoverPoint KVSS is a command-line utility that can apply bookmarks to Windows-based applications—for example, Microsoft SQL Server and Exchange Server.

- The application supports Microsoft **Volume Shadow Copy Service** (**VSCS**).

- It does not have any storage array hardware dependencies for RP4VM.

The next section will discuss the local replication and remote replication architectures of RP4VM on VxRail.

Local replication

If you plan to enable the CDP feature on VxRail locally, you can enable local protection with VxRail. *Figure 9.6* shows the local replication architecture of RP4VM on VxRail:

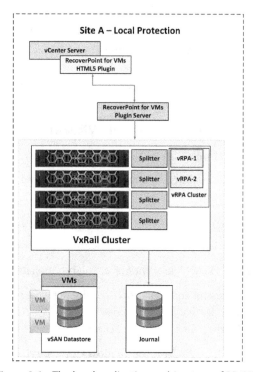

Figure 9.6 – The local replication architecture of RP4VM

It includes the following core components:

- One VMware vCenter Server appliance and the RP4VM HTML5 plugin.

- One RP4VM plugin server.

- One VxRail cluster with four nodes; a splitter is enabled on each VxRail node.

- One RecoverPoint appliance cluster has two vRPA nodes (**vRPA-1** and **vRPA-2**), which are running on two VxRail nodes. Each RecoverPoint appliance cluster contains, at a minimum, two vRPAs.

- All production VMs are stored in the vSAN datastore.

- All PIT images of VMs are stored in the journal volume, and they can be SAN, NAS, FC, or iSCSI volumes.

In *Figure 9.6*, you learned the local replication architecture of RP4VM in a VxRail cluster.

Remote replication

If you plan to enable the CDP feature on VxRail across two sites, you can enable remote protection with VxRail. *Figure 9.7* shows the remote replication architecture of RP4VM on VxRail:

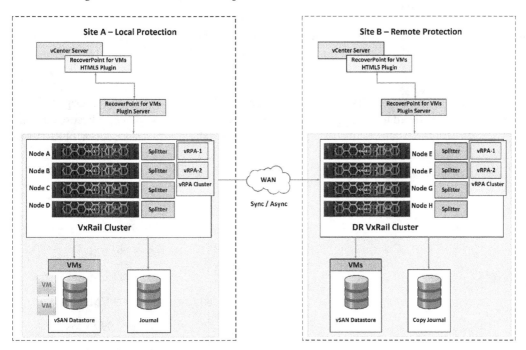

Figure 9.7 – The remote replication architecture of RP4VM

It includes the following core components:

- Two VMware vCenter Server appliances containing the RP4VM HTML5 plugin.

- Two RP4VM plugin servers.

- One VxRail cluster with four nodes; a splitter is enabled on each VxRail node at **Site A** and **Site B**.

- At each site, there is one RecoverPoint appliance cluster with two vRPA nodes (**vRPA-1** and **vRPA-2**), running on two VxRail nodes. Each RecoverPoint appliance cluster contains, at a minimum, two vRPAs.

- All production VMs are stored in the vSAN datastore.

- The WAN connection is used for synchronous and asynchronous replication between **Site A** and **Site B**.

- All PIT images of VMs are stored in the journal volume, and they can be SAN, NAS, FC, or iSCSI volumes.

- All PIT copy images of VMs at **Site A** are stored in this volume, and they can be SAN, NAS, FC, or iSCSI volumes.

In *Figure 9.7*, you learned the remote replication architecture of RP4VM in a VxRail cluster.

Licensing

Starting from the RP4VM 5.2 release, licensing is offered based on the socket and subscription, and the original per-VM tiered licensing is still available. Now, you can choose different license options based on your business requirements. *Table 9.3* provides the details of each license option of RP4VM:

Licensing Option	Description	Details
Per VM	Supports the number of production VMs you need to protect.	License tier options: • 15 to 99 VMs • 100 to 249 VMs • 250 to 449 VMs • 500 to 999 VMs • 1,000 to 1,999 VMs • Over 2,000 VMs
CPU socket-based licensing	Counts the number of physical CPU sockets in the ESXi platform hosting production VMs.	If both per-VM and per-CPU socket licenses are installed, RP4VM automatically converts to a per-CPU socket license at a conversion rate of 15 VMs per socket.
Subscription-based licensing	It can offer three license options, per-VM or per-CPU socket.	License tier options: • 1-year subscription, which allows you to get the benefit of pay-as-you-go flexibility • 1-year subscription and 3-year commitment • 3-year pre-paid subscription

Table 9.3 – The license option of RP4VM

Simple support matrix

When you design the setup of RP4VM on VxRail, you need to verify the software compatibility of RP4VM by using a simple support matrix with the core components, including the VxRail software release, the VMware vSphere release, the vRPA version, and the RP4VM HTML5 plugin server. *Table 9.4* shows details of the supported VxRail platforms:

VxRail Platform	VMware Platform	Minimum Supported vRPA Version
VxRail 7.0.xxx	VMware vSphere 7.0, Update 1, Update 2, and Update 3	RP4VM 5.3
VxRail 4.7.xxx	VMware vSphere 6.7, Update 1, Update 2, and Update 3	RP4VM 5.3
VxRail 4.5.xxx	VMware vSphere 6.5, Update 1, Update 2, and Update 3	RP4VM 5.3

Table 9.4 – The supported VxRail platforms

> **Important note**
> Upgrading to RP4VM 5.3 SP3 is only supported from version 5.2 and later. The HTML5 plugin is supported in vSphere 6.7 Update 1 and later. The vSphere Web Client plugin cannot be used in vCenter 6.5 Update 3n or 6.7 Update 3l and later.

The RP4VM HTML5 plugin server also supports the different vRPA versions. *Table 9.5* provides details on the supported plugin servers:

vRPA	Minimum Supported Plugin Versions
vRPA 5.3	RecoverPoint plugin server version 5.3.22
vRPA 5.3.1	RecoverPoint plugin server version 5.3.1.420
vRPA 5.3.1.1	RecoverPoint plugin server version 5.3.1.1.11
vRPA 5.3.2.x	RecoverPoint plugin server version 5.3.2.119
vRPA 5.3.3	RecoverPoint plugin server version 5.3.2.119

Table 9.5 – The supported plugin servers

When you execute the failover or failback operation of RP4VM on a VxRail cluster, you may need to change the IP address of protected VMs. *Table 9.6* shows the supported guest operating systems for changing the IP address feature:

Operating System	Supported Versions
Microsoft Windows Server	Microsoft Windows 8 and 10 Microsoft Windows 2008 R2, 2012, 2016, 2019, and 2022
Red Hat Enterprise Linux (RHEL) Server	RHEL 6.5, 7.1, 7.2, 7.5, 7.6, and 8.2
Ubuntu Server	Ubuntu 15.10

Table 9.6 – The supported guest operating systems for changing the IP address feature

The next section will discuss the benefits of RP4VM.

The benefits of RP4VM

When you enable RP4VM on VxRail, it can provide the following benefits:

- It enables data center migration with minor interruptions.
- The RPO settings can configure to be 0 seconds.
- A single RecoverPoint appliance can support multi-vSphere cluster protection.
- It supports any type of PIT recovery of VMs.
- It does not have any storage array hardware dependencies for RP4VM, and it can support SAN, NFS, DAS, FC, iSCSI, and vSAN.
- It provides synchronous and asynchronous concurrent replication to local and remote copies of the VMs.
- RP4VM is fully integrated with VMware vCenter Server.
- It allows continuous data replication without VMware snapshots locally and remotely.
- It provides DR and operational recovery with VM-level granularity.
- There is no performance impact on the application during data replication.
- It can support different cloud vendors—for example, **Amazon Web Services (AWS)**, **VMware Cloud (VMC)**, Microsoft Azure, and **Google Cloud Platform (GCP)**.
- It can support VMware NSX and NSX-T solutions.
- It can support VMware vRealize Operations Manager.

- It can fully support **Representational State Transfer (REST) application programming interfaces (APIs)**.

- It can deliver recovery operations, including failover and fallback tests.

The next section will discuss the design of RP4VM on VxRail.

The design of RP4VM on VxRail

This section will discuss the low-level design of RP4VM on VxRail, including local replication and remote replication. *Figure 9.8* shows a local replication configuration of RP4VM on VxRail:

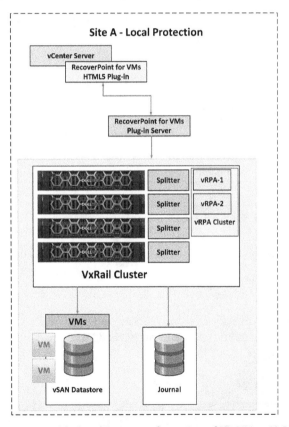

Figure 9.8 – A local replication configuration of RP4VM on VxRail

In this configuration, it is a standard VxRail cluster with four nodes. *Table 9.7* shows the hardware configuration of each VxRail node. Each P670F installed one 800 GB SSD (cache tier), four 7.68 TB **SSDs** (capacity tier), one quad-port 10 GB Ethernet adapter, and one PCIe dual-port 10 GB Ethernet adapter:

VxRail Model	VxRail P670F
CPU Model	2 x Intel Xeon Gold 5317 3G, 12C/24T
Memory	512 GB (8 x 64 GB)
Network Adapter 1	Intel Ethernet X710 Quad Port 10GbE SFP+, OCP NIC 3
Network Adapter 2	Intel X710 Dual Port 10GbE SFP+ Adapter, PCIe low profile
Cache Drive	1 x 800 GB SSD SAS drive
Capacity Drive	4 x 7.68 TB SSD SAS drive

Table 9.7 – A sample configuration of VxRail P670F

Each VxRail P670F includes the following software; *Table 9.8* shows the software edition of each VxRail, VMware component, and RP4VM:

VxRail Software Release	7.0.370 build 27485531
VMware ESXi Edition	7.0 Update 3d
VMware vCenter Server	7.0 Update 3d
VMware vSAN	7.0 Update 3d
vRPA	5.3.3
RP4VM Plugin Server	5.3.3

Table 9.8 – The VxRail software release

The next section will discuss the network design of the scenario presented in *Figure 9.8*.

Network settings

Each VxRail P670F has six 10 Gb network ports, as shown in *Figure 9.9*. **P1** and **P2** are used for ESXi and VxRail management networks and witness traffic. **P3** and **P4** are used for the vSAN and vMotion networks. **P5** and **P6** are used for RP4VM:

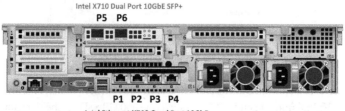

Figure 9.9 – Rear view of VxRail P670F

For the network design of the VxRail cluster, you can refer to the following table:

Network Traffic	NIOC Shares	VMkernel Ports	VLAN	P1	P2	P3	P4
Management network	40%	vmk2	101	Standby	Active	Unused	Unused
vCenter Server management network	N/A	N/A	101	Standby	Active	Unused	Unused
VxRail management network	N/A	vmk1	101	Standby	Active	Unused	Unused
vSAN network	100%	vmk3	200	Unused	Unused	Active	Standby
vMotion network	50%	vmk4	100	Unused	Unused	Standby	Active
VMs	60%	N/A	N/A	Active	Standby	Unused	Unused

Table 9.9 – The network layout of VxRail

For the network design of RP4VM, you can refer to the following table:

Network Traffic	NIOC Shares	VMkernel Ports	VLAN	P5	P6
WAN and LAN network of RP4VM	N/A	vmk5	101	Active	Standby
Data network of RP4VM	N/A	vmk6	400	Standby	Active

Table 9.10 – The network layout of RP4VM

Tables 9.9 and *9.10* help you understand the network layout of RP4VM on VxRail. The next section will discuss the storage design of the scenario presented in *Figure 9.8*.

Storage settings

The VxRail P670F installed one 800 GB SSD and four 7.68 TB SSDs; you can create a vSAN disk group with one 800 GB SSD for the cache tier and four 7.68 TB SSDs for the capacity tier. For the disk groups upgrade, you can refer to the *Design of disk groups on VxRail P-Series* section in *Chapter 5*. Since P670F is an All-Flash model, it can support RAID-1, RAID-5, and RAID-6 site protection. The next section will discuss the required software licenses for this scenario.

Software licenses

If stretched clusters are being deployed in this scenario, they require the following VMware licenses. *Table 9.11* shows a summary of all bundled software licenses on VxRail P670F:

Software Name	License Edition	Quantity	Remark
VMware vSphere	VMware vSphere Enterprise Plus per CPU	8	N/A
VMware vSAN	VMware vSAN Enterprise per CPU	8	You can choose the vSAN Enterprise or Enterprise Plus edition.
VMware vCenter Server	VMware vCenter Server Standard instance	1	I suggest deploying the customer-supplied vCenter Server; it requires an optional vCenter Server Standard license.
VMware vRealize Log Insight	Bundled with VxRail Appliance	1	N/A
VMware Replication	Bundled with VxRail Appliance	N/A	N/A
Dell EMC RP4VM	Bundled with VxRail Appliance	N/A	You can order per-VM licenses or per-CPU socket licenses if you need more VM licenses.

Table 9.11 – A summary of all bundled software licenses on VxRail P670F

With the preceding information, you now understand the low-level design of local replication for RP4VM, as depicted in *Figure 9.8*.

Now, we will discuss the remote replication configuration of RP4VM on VxRail, as shown in *Figure 9.10*:

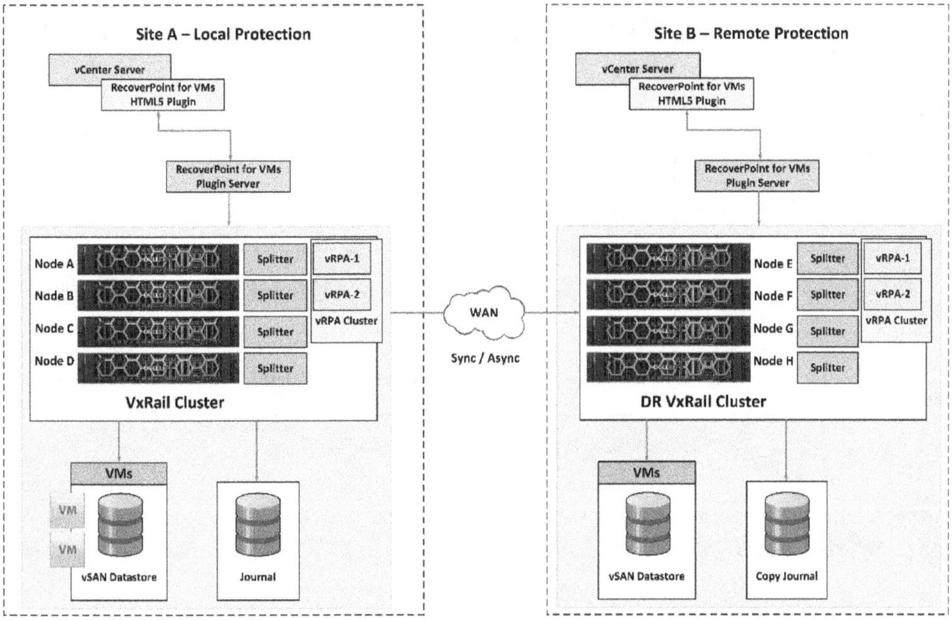

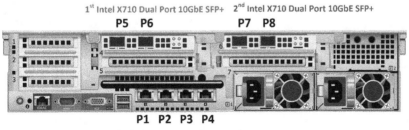

Figure 9.10 – A remote replication configuration of RP4VM on VxRail

This configuration shows a standard VxRail cluster with four nodes installed in **Site A** and **Site B**. Each P670F installed one 800 GB SSD (cache tier), four 7.68 TB SSDs (capacity tier), one quad-port 10 Gb Ethernet adapter, and two PCIe dual-port 10 Gb Ethernet adapters. For storage, software, and VxRail network requirements, you can refer to the low-level design of local replication in *Figure 9.8*.

In terms of the network settings, each VxRail P670F has eight 10 Gb network ports, as shown in *Figure 9.11*. **P1** and **P2** are used for ESXi and the VxRail management network and witness traffic. **P3** and **P4** are used for the vSAN and vMotion networks. **P5** and **P6** are used for the LAN and WAN networks of RP4VM, and **P7** and **P8** are used for the data network of RP4VM:

Figure 9.11 – Rear view of VxRail P670F

Table 9.12 shows the network layout used for RP4VM:

Network Traffic	NIOC Shares	VMkernel Ports	VLAN	P5	P6	P7	P8
WAN network of RP4VM	N/A	vmk5	102	Active	Standby	Unused	Unused
LAN network of RP4VM	N/A	vmk6	101	Standby	Active	Unused	Unused
Data network of RP4VM	N/A	vmk7	400	Unused	Unused	Active	Standby
Data network of RP4VM	N/A	vmk8	400	Unused	Unused	Standby	Active

Table 9.12 – The network layout of RP4VM

With the preceding two scenarios, you learned about the low-level design of local replication for RP4VM, depicted in *Figures 9.8* and *9.10*. The next section will discuss some failover scenarios of RP4VM.

Failover scenarios of RP4VM on VxRail

This section will discuss some failover scenarios of RP4VM, including a faulty vRPA, a faulty journal, replication links disconnected, a faulty plugin server, and a faulty vCenter Server instance.

Failure scenario 1

Figure 9.12 shows local replication of RP4VM. Which status will the protected VMs trigger if **vRPA-1** is faulty in the vRPA cluster?

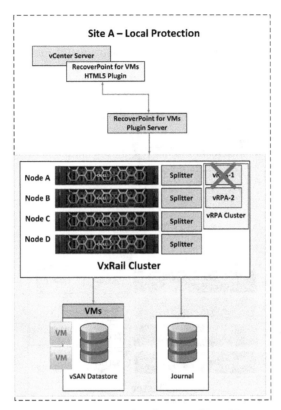

Figure 9.12 – Local replication of RP4VM

The running VMs can still be running in the VxRail cluster, and the services of **vRPA-1** will fail over to **vRPA-2**. The location replication can continue to execute. You can still manage all operational tasks of RP4VM within vCenter Server. The next section will discuss another failure scenario.

Failure scenario 2

Figure 9.13 shows local replication of RP4VM. Which status will the protected VMs trigger if the connection of journal volumes is disconnected?

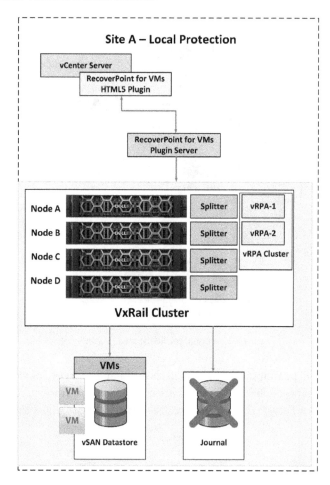

Figure 9.13 – Local replication of RP4VM

When the connection of journal volumes between RP4VM is disconnected, all local replication tasks will be faulty. If the vRPA clusters cannot connect to the journal volumes, you will lose all PIT images of VMs. The next section will discuss another failure scenario.

Failure scenario 3

Figure 9.14 shows local and remote replication of RP4VM. Which status will the protected VMs trigger if replication links are disconnected across two sites?

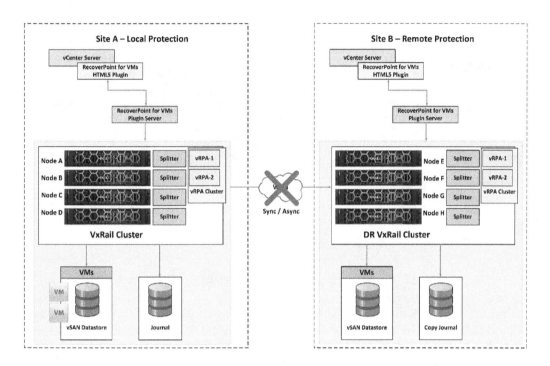

Figure 9.14 – Local and remote replication of RP4VM

In this configuration, the remote replication sessions will be stopped if replication links are disconnected across two sites. Local replication sessions are not impacted by this issue. The remote replication sessions will resume automatically and synchronize the delta change of data to **Copy Journal** when the replication links are connected across two sites. The next section will discuss another failure scenario.

Failure scenario 4

Figure 9.15 shows local and remote replication of RP4VM. Which status will the protected VMs trigger if **RecoverPoint for VMs Plugin Server** is faulty at **Site A**?

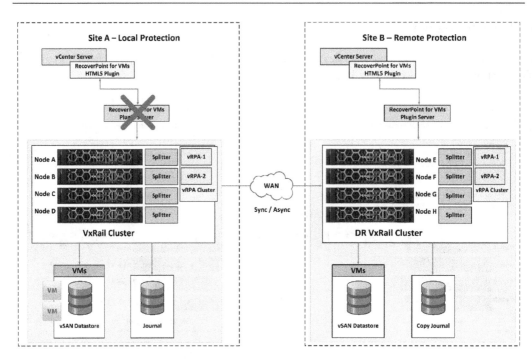

Figure 9.15 – Local and remote replication architecture of RP4VM

In this configuration, if **RecoverPoint for VMs Plugin Server** is faulty at **Site A**, it will impact the local and remote replication sessions. You cannot manage all replication tasks within **vCenter Server** because **vCenter Server** contains two different certificates; the first certificate is by the vRPAs, which validate it and then send it to **RecoverPoint for VMs Plugin Server**. **RecoverPoint for VMs Plugin Server** then attempts to contact **vCenter Server** to validate that the certificates match. The next section will discuss another failure scenario.

Failure scenario 5

Figure 9.16 shows local and remote replication of RP4VM. Which status will the protected VMs trigger if **vCenter Server** is faulty at **Site A**?

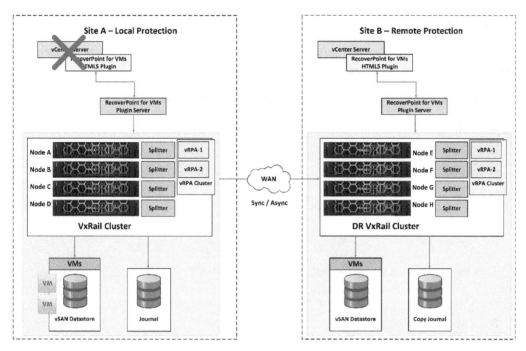

Figure 9.16 – The remote replication architecture of RP4VM

In this configuration, if **vCenter Server** is faulty at **Site A**, it will impact the local and remote replication sessions. The reason is the same as *Failure scenario 4*.

With these scenarios, you should now understand the expected results of failure scenarios on RP4VM on VxRail.

Summary

In this chapter, we covered an overview and the design of RP4VM on VxRail, including the network, hardware, and software requirements, and some failure scenarios. When you plan to enable CDP features and need fast data recovery, RP4VM is a good option.

In the following, and last, chapter, you will learn about the design of VxRail with Veeam Backup & Replication and the best practices of this solution.

Questions

The following is a short list of review questions to help reinforce your learning and help you identify areas which require some improvement.

1. How many configurations of RecoverPoint appliance adapters are there on RP4VM?

    A. One

    B. Two

    C. Three

    D. Four

    E. Five

    F. Six

2. How many performance profiles are supported by RP4VM?

    A. One

    B. Two

    C. Three

    D. Four

    E. Five

    F. Six

3. Which of the following are supported performance profiles on RP4VM?

    A. **Bronze profile**: Two vCPUs and 8 GB of memory

    B. **Bronze profile**: Two vCPUs and 6 GB of memory

    C. **Silver profile**: Four vCPUs and 8 GB of memory

    D. **Silver profile**: Four vCPUs and 10 GB of memory

    E. **Gold profile**: Eight vCPUs and 8 GB of memory

    F. **Gold profile**: Eight vCPUs and 10 GB of memory

4. How many vRPAs can a single RP4VM CG be assigned to?

    A. One

    B. Two

    C. Three

    D. Four

E.   Five

F.   Six

5.   If an application requires four VMs, how can the application have consistent PIT images using RP4VM?

A.   Place all VMs in the same CG

B.   Place two VMs in their own CG

C.   Place each VM in its own CG

D.   Create two CGs and assign two VMs to each one

E.   Place three VMs in the same CG

F.   Create three CGs only

6.   Which of the following replication modes are supported by RP4VM?

A.   Local replication

B.   Mirror replication

C.   RAID-5 protection

D.   RAID-1 protection

E.   Remote replication

F.   RAID-6 protection

7.   Which of the following license options are supported by RP4VM?

A.   Number of protected VMs

B.   Number of protected ESXi hosts

C.   Number of CPU socket-based licenses

D.   Subscription-based licenses

E.   Number of CPU core-based licenses

F.   All of these

8.   Which configuration of RecoverPoint appliance adapters for RP4VM can provide the best performance and HA?

A.   One adapter

B.   Two adapters

C.   Three adapters

D.   Four adapters

E. Five adapters

F. Six adapters

9. Which configuration of RecoverPoint appliance adapters for RP4VM cannot provide HA?

A. One adapter

B. Two adapters

C. Three adapters

D. Four adapters

E. Five adapters

F. Six adapters

10. Which of the following failure scenarios can trigger the VM's replication task to stop?

A.

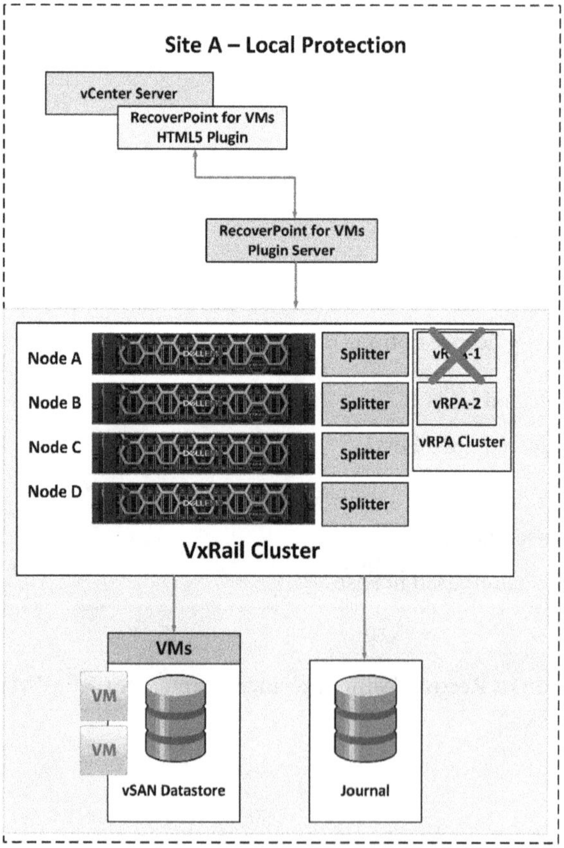

Figure 9.17 – Local replication of RP4VM

B.

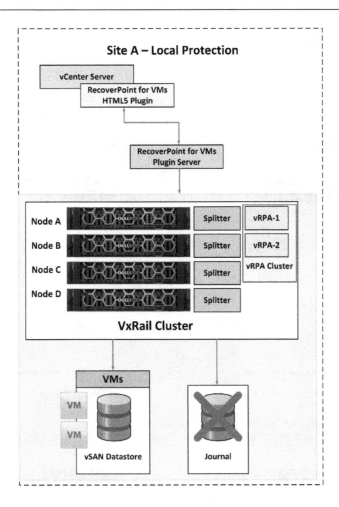

Figure 9.18 – Local replication of RP4VM

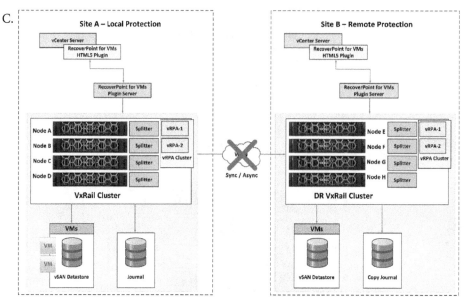

Figure 9.19 – Remote replication of RP4VM

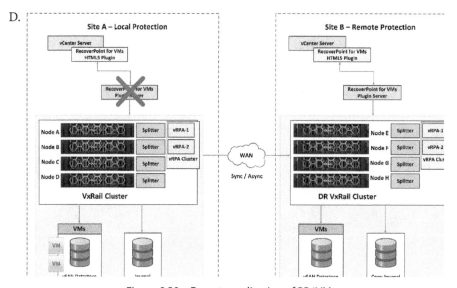

Figure 9.20 – Remote replication of RP4VM

E.

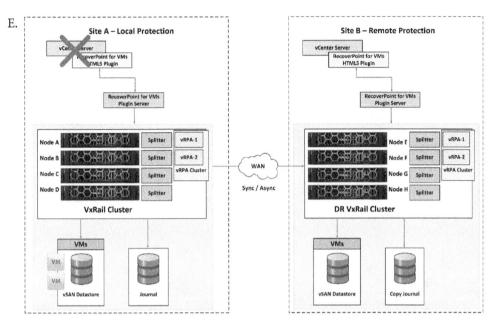

Figure 9.21 – Remote replication of RP4VM

11. What is the maximum number of vRPAs that can be supported in a vRPA cluster?

    A. Four

    B. Five

    C. Six

    D. Seven

    E. Eight

    F. Nine

12. Which service must be installed on each ESXi host/VxRail node during the deployment of RP4VM?

    A. Journal volume

    B. Copy journal volume

    C. CG

    D. RP4VM splitter

    E. RP4VM plugin server

    F. VMware vCenter Server

10
Design of VxRail with Veeam Backup

In the previous chapter, we covered an overview and the design of **RecoverPoint for Virtual Machines** (**RP4VMs**) on VxRail, including network, hardware, and software requirements, and some failure scenarios. When you plan to enable **Continuous Data Protection** (**CDP**) features and fast data recovery, RP4VMs is a good option.

Previous chapters of this book discussed the Dell and VMware features on a VxRail cluster, such as VxRail cluster expansion, a VMware stretched cluster, and RP4VMs. This chapter will discuss the third-party backup and recovery software **Veeam Backup and Replication** (**VBR**), which can provide data protection for **virtual machines** (**VMs**) and physical machines. You will see an overview and the benefits of VBR and learn about designing integration with VxRail and VBR.

This final chapter of the book covers the following main topics:

- Overview of VBR
- Overview of VxRail with VBR
- Benefits of VxRail with VBR
- Design of VxRail with VBR
- Recovery scenarios of VxRail with VBR

Overview of VBR

VBR is data protection software that can deliver ransomware protection and data recovery for your enterprise data, including data stored on VMware vSphere, Microsoft Hyper-V, Nutanix **Acropolis Hypervisor** (**AHV**), and cloud platforms (**Amazon Web Services** (**AWS**), Microsoft Azure, and **Google Cloud Platform** (**GCP**)). VBR contains the following components:

- **Backup server**: This is a Windows-based physical machine or a VM on VBR. It is a core component of the Veeam backup infrastructure. This backup server can carry out all types of data recovery tasks, including backup creation, data replication, and recovery verification.

- **Backup proxy server**: This is an architecture component in the Veeam backup infrastructure. When the backup server executes backup tasks, the backup proxy server processes backup jobs and delivers backup traffic. The backup proxy server can carry out different tasks, including the retrieval of VM data from storage, and performing compression, deduplication, encryption, and so on.

- **Backup repository**: This is a storage pool where all backup files are kept, including VM copies and metadata for replicated VMs. The backup repository supports different storage types, including direct attached storage, **Common Internet File System** (**CIFS**), deduplication storage appliances, and object storage.

- **Gateway server**: This is used to make a connection between the backup server and the backup repository. The gateway server is required when you deploy these types of backup repositories, including a Dell EMC Data Domain storage appliance, an HPE StoreOnce storage appliance, and shared folder backup repositories.

- **VBR console**: This is a management console that connects to the VBR server. By default, the VBR console is installed on the backup server.

This section will discuss the following topics:

- The architecture of VBR
- The 3-2-1-1-0 backup rule
- Overview of Veeam backup job modes
- Overview of Veeam Agent backup

The next subsection will discuss which machines can be protected with VBR.

Which platforms can be protected?

The latest version of VBR is v11a, which was released in July 2022. It supports data protection for the following platforms:

- VMware vSphere and vCenter Server
- Microsoft Hyper-V and **System Center Virtual Machine Manager (SCVMM)**
- Nutanix AHV and Prism
- **Network-Attached Storage (NAS)**, including **Network File System (NFS)**, **Server Message Block (SMB)**, and Windows- or Linux-based file servers
- Microsoft Windows, Linux, Unix, and Mac (in Veeam Agent)

Figure 10.1 shows the architecture of a simple deployment of VBR. There is a VBR server that contains three components: a **backup proxy server**, a **backup repository**, and a **VBR console**. The **VMs** on three VMware ESXi hosts are protected by a VBR server. This is a simple deployment of VBR; all of VBR's components and database are running on a single machine. This deployment is suitable for a small-scale VMware environment:

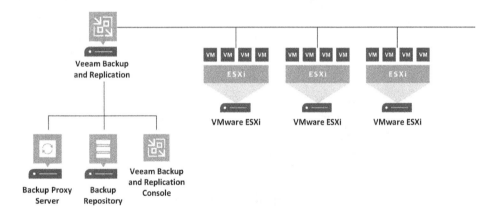

Figure 10.1 – A simple deployment of VBR

Now, we will discuss the backup flow in a simple deployment of VBR. The following is the backup flow, also illustrated in *Figure 10.2*:

1. When you start a new backup job, VBR starts the Veeam Backup Enterprise Manager service. All backup job settings are read from the configuration database. Then, Veeam Backup Manager connects to the Veeam backup service and assigns the backup proxy server and backup repository to execute the backup jobs.

2. Veeam Backup Manager makes a connection with the Veeam Data Mover service on the backup repository and backup proxy server.

3. The data transfer between the backup repository and the backup proxy server is carried out.

4. VBR requests vCenter Server or the ESXi host to create a VM snapshot.

5. Veeam Data Mover reads the VM data from the read-only VM (VM snapshot) and transfers the data to the backup repository.

6. After the backup proxy server finishes reading VM data, VBR requests for vCenter Server or the ESXi host to remove the VM snapshot:

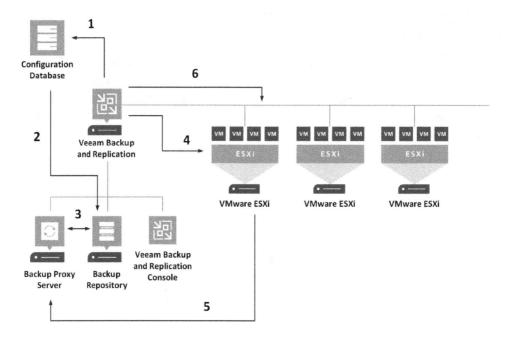

Figure 10.2 – The backup flow in VBR

Figure 10.3 shows the architecture of an advanced deployment of VBR. There is a VBR server with a console, and a database server, which can be a physical machine or a VM. The backup repository is a storage server or storage appliance where all backup files are kept, including the VM copies and metadata for replicated VMs. The backup proxy server can be a physical machine or a VM. This is the advanced deployment of VBR; all VBR components are separated into different hosts. This deployment is suitable for a large-scale VMware environment:

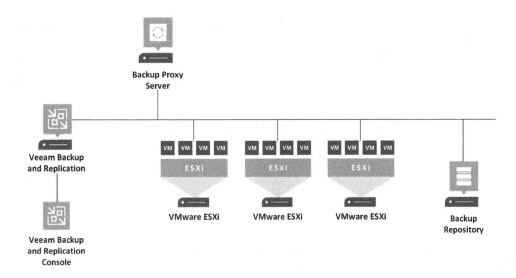

Figure 10.3 – The advanced deployment of VBR

Now, we will discuss the backup flow in the advanced deployment of VBR. The following is the backup flow, also illustrated in *Figure 10.4*:

1. When you start a new backup job, VBR starts the Veeam Backup Enterprise Manager service. All backup job settings are read from the configuration database. Then, Veeam Backup Manager connects to the Veeam backup service and assigns the backup proxy server and backup repository to execute the backup jobs.

2. Veeam Backup Manager makes a connection with the Veeam Data Mover service on the backup repository and backup proxy server.

3. The data transfer between the backup repository and the backup proxy server is carried out.

4. VBR requests vCenter Server or the ESXi host to create a VM snapshot.

5. Veeam Data Mover reads the VM data from the read-only VM (VM snapshot) and transfers the data to the backup repository.

6.   After the backup proxy server finishes reading VM data, VBR requests for vCenter Server or the ESXi host to remove the VM snapshot:

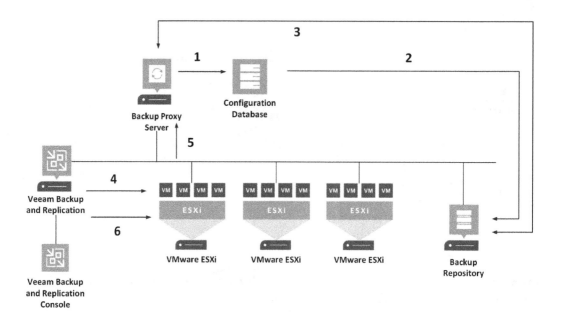

Figure 10.4 – The backup flow in VBR

Figure 10.5 shows the architecture of VBR. There are two locations: the **production site** and the **remote site**. In the production site, there are four core components: the VBR server, the backup proxy server, the backup repository, and WAN Accelerator. WAN Accelerator is a Veeam technology that optimizes the transfer of data to remote locations between WAN connections. You can enable this feature when you need to set up off-site backup copy jobs and VM replication. The VMs are replicated to the VMware ESXi host from the production site to the remote site:

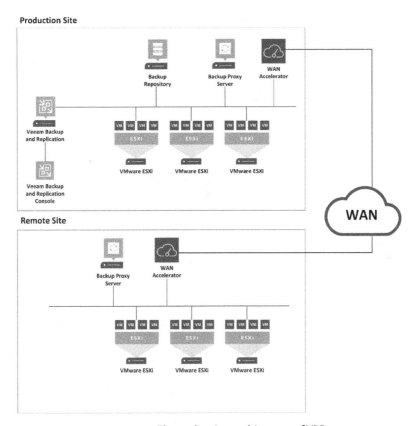

Figure 10.5 – The replication architecture of VBR

Now, we will discuss the VM replication flow in VBR. The following is the replication flow between the two sites, also illustrated in *Figure 10.6*:

1. When you start a new replication job, VBR starts the Veeam Backup Manager service. All replication job settings are read from the configuration database. Then, Veeam Backup Enterprise Manager connects to the Veeam backup service and selects a backup proxy server and backup repository to execute the replication jobs.

2. The VBR server assigns the backup proxy server and backup repository to execute the replication tasks.

3. The source backup proxy server makes a connection with the target backup proxy server and backup repository. VBR compresses the VM data (for replication) through WAN Accelerator and copies it to the target site.

4. VBR requests for vCenter Server or the ESXi host to create a VM snapshot. The VM disks are put in read-only mode, and all virtual disks receive a delta file.

5. The source backup proxy server reads the VM data from the read-only VM disks, and then each change to the VM during replication is written to delta files.

6. The target backup proxy server decompresses VM data and writes the data to the destination datastore.

7. After the backup proxy server finishes reading VM data, VBR requests for vCenter Server or the ESXi host to remove the VM snapshot:

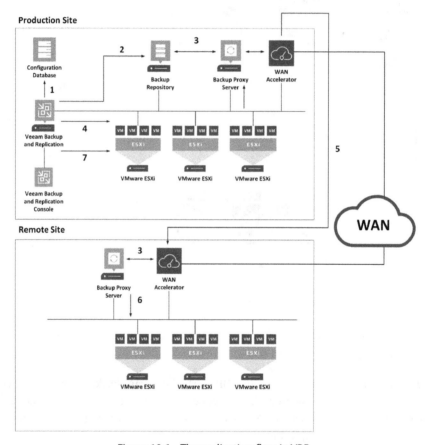

Figure 10.6 – The replication flow in VBR

After going through the preceding scenarios, you have got an overview and understood the backup flow of VBR. We will discuss the 3-2-1-1-0 backup rule in the next subsection.

The 3-2-1-1-0 backup rule

To protect your data, you should ensure your data is stored safely. In this section, we will discuss the **3-2-1-1-0** backup rule, which denotes the following:

- **3**: Three different copies of data. You should keep three copies of your data—one is the primary copy (production), and two are the secondary copies (local backup and remote or off-site backup).

- **2**: Two different types of media. You should store your backup data in at least two different types of media.

- **1**: One off-site copy. You must keep at least one copy of your data in a secure off-site location.

- **1**: One copy offline. You should keep at least one copy of your data offline.

- **0**: Zero errors. Make sure you have a verified backup without errors. You must ensure your backup data can be recovered with zero errors.

Figure 10.7 also provides an explanation of the 3-2-1-1-0 backup rule. This backup rule is recognized as the best practice by Veeam. If we follow this backup rule, we can minimize the risk of data loss and a **single point of failure** (**SPOF**):

Figure 10.7 – The 3-2-1-1-0 backup rule (copyright from Veeam)

Licensing

Veeam software contains two license options, including **Veeam Availability Suite** and VBR. Both options are part of the **Veeam Universal License** (**VUL**), which is available as a subscription or perpetual option. *Table 10.1* shows a comparison of Veeam licenses:

Supported Features	Veeam Availability Suite	VBR
Virtual platform, including VMware vSphere, Microsoft Hyper-V, and Nutanix AHV	Supported	Supported
Physical platform, including Microsoft Windows, Linux, AIX, Oracle Solaris, Mac, and NAS	Supported	Supported
Cloud platform, including AWS, Azure, and GCP native backup and recovery	Supported	Supported
Application support, including Microsoft SQL, Exchange, Oracle, MySQL, PostgreSQL, and SAP	Supported	Supported
Advanced application recovery on Veeam Explorer	Supported	Supported
Veeam One	Supported	Not supported
VUL	Supported	Supported
Veeam Backup for Microsoft 365	Licensed separately	Licensed separately
Veeam Disaster Recovery Orchestrator	Optional add-on license	Optional add-on license

Table 10.1 – Veeam license comparison

VUL is a portal license that supports moving easily across different workloads—for example, VMware vSphere, Microsoft Hyper-V, Windows, and cloud platforms. *Table 10.2* shows a summary of VUL's supported workloads:

Example Workloads	Details
Cloud platform	Microsoft Azure, GCP, AWS, IBM Cloud, Kubernetes
Virtual platform	VMware vSphere, Microsoft Hyper, Nutanix AHV, Red Hat Virtualization
Physical platform	Microsoft Windows, Linux, Unix, macOS, NAS
Applications	Microsoft Exchange and SQL, Oracle, SAP Hana, PostgreSQL, Kubernetes
Software as a Service (SaaS)	Microsoft 365, Microsoft Teams, Salesforce
Unstructured data	NAS, SharePoint

Table 10.2 – A summary of VUL's supported workloads

> **Important note**
> Kubernetes, Microsoft 365 (including Teams), and Salesforce require a separate license.

Overview of backup job modes

VBR v11 contains different backup modes, including reverse incremental, forever forward incremental, and forward incremental. Now, we will discuss each backup option.

Figure 10.8 shows the backup chain for reverse incremental backup. This backup methodology contains the last full backup file (**vbk**) and a set of reverse incremental backup files (**vrb**). For example, you can configure the backup as follows. If you choose this backup mode, it will add all new restore points to the backup chain and rebuild the full backup file:

- The backup job starts on Sunday.

- The backup method is configured in reverse incremental mode.

- The retention policy is configured to six restore points:

Figure 10.8 – Reverse incremental backup

Figure 10.9 shows the backup chain for forever forward incremental backup; this backup methodology contains the first full backup file (**vbk**) and a set of forward incremental backup files (**vrb**):

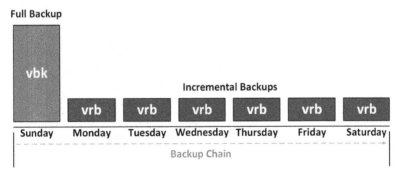

Figure 10.9 – Forever forward incremental

Veeam Agent backup

If the backup job of your workload cannot create VM snapshots or storage snapshots by VBR, you can choose the Veeam Agent backup. VBR also supports the data backup of physical machines, including Microsoft Windows, Linux, Unix, and Mac. You can use VBR to manage Veeam Agent for physical machines. There is no Veeam proxy server when you execute the Veeam Agent backup—it is recommended to use a separate NIC using a backup network so that it does not saturate production traffic. The only exception is when the backup from the storage snapshots feature is enabled. *Figure 10.10* shows the architecture of the Veeam Agent backup:

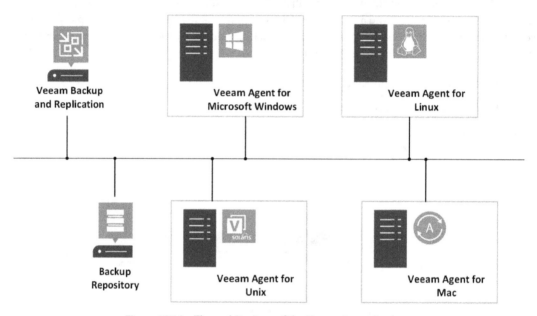

Figure 10.10 – The architecture of the Veeam Agent backup

Table 10.3 shows the supported features of Veeam Agent on Windows, Linux, and Mac:

	Veeam Agent for Windows	Veeam Agent for Linux	Veeam Agent for Mac
Backup modes	Volume-level backup, file-level backup, and the entire machine	Volume-level backup, file-level backup, and the entire machine	User data

Destination	Local storage, shared folder, and Veeam backup repository	Local storage and shared folder	Local storage, shared folder, and Veeam backup repository
Application-aware processing	Active Directory, Exchange, SQL, Oracle, and SharePoint	Oracle, PostgreSQL, and MySQL	N/A
Custom recovery media	Supported	Supported	N/A
Built-in snapshot and **changed block tracking (CBT)**	Supported	Supported	N/A
File indexing	Supported	Supported	N/A
Data encryption	Supported	Supported	Supported
Backup cache	Supported	N/A	N/A
Transaction log processing for Microsoft SQL and Oracle	Supported	N/A	N/A

Table 10.3 – A comparison of Veeam Agent for Windows, Linux, and Mac

The next section will discuss an overview of VxRail with VBR.

Overview of VxRail with VBR

In previous chapters of this book, we mentioned that VxRail bundles some data protection software, such as VMware vSphere Replication, vSAN stretched clusters, and Dell EMC RP4VMs. If you do not use these data protection features on a VxRail cluster, you can use third-party data backup and recovery software—for example, VBR. *Figure 10.11* shows a sample architecture of VxRail with VBR:

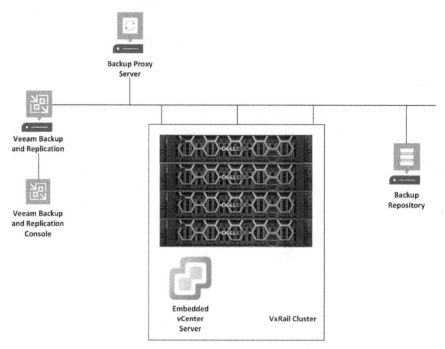

Figure 10.11 – A sample architecture of VxRail with VBR

In the configuration shown in *Figure 10.11*, there are four Veeam components—the VBR server, the VBR console, the backup proxy server, and the backup repository one VxRail cluster with four nodes, and one embedded vCenter Server instance. We can create a backup job to protect the VMs on the VxRail cluster with the VBR server. All backup images of VMs will be stored in the backup repository. In this scenario, the backup proxy server is a virtual host that can be deployed into the VxRail cluster. The VBR server is a VM that runs outside of the VxRail cluster. *Table 10.4* shows a summary of each component in *Figure 10.11*:

Components	Details
VBR server	A backup server is a VM that is installed outside of the VxRail cluster.
VBR console	This is the management console of the VBR server that is installed on VBR.
Backup proxy server	This is a VM that is installed on the VxRail cluster.
Backup repository	This is SAN storage or a local storage server.
VxRail cluster	VxRail All-Flash, Hybrid, or NVMe nodes.
VMware vCenter Server	VxRail embedded vCenter Server instance.

Table 10.4 – A summary of each component in Figure 10.11

Now, we will discuss the other scenario of VxRail being used with VBR: as a **disaster recovery (DR)** solution across two sites.

Figure 10.12 shows a sample architecture of VxRail with VBR across two sites:

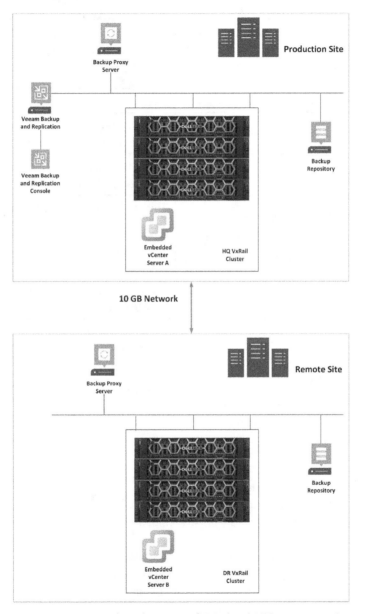

Figure 10.12 – A sample architecture of VxRail with VBR across two sites

In the configuration in *Figure 10.12*, there are six Veeam components, including the VBR server, the VBR console, two backup proxy servers, two backup repositories, one VxRail cluster with four nodes, and one embedded vCenter Server instance per site. We can create VM replication jobs and replicate the VMs from the **production site** to the **remote site**. All replication images of VMs will be stored in the backup repository at the remote site. In this scenario, the backup proxy server is a virtual host that can be deployed into the VxRail cluster. The VBR server is a VM that runs outside of the VxRail cluster. The network traffic of the replication link is 10 GB between the two sites. *Table 10.5* shows a summary of each component in *Figure 10.12*:

Sites	Components	Details
Production site	VBR server	A backup server is a VM that is installed outside of the VxRail cluster.
	VBR console	This is the management console of the VBR server that is installed on VBR.
	Backup proxy server	This is a VM that is installed on the VxRail cluster.
	Backup repository	This is SAN storage or a local storage server.
	VxRail cluster	VxRail All-Flash, Hybrid, or NVMe nodes.
	VMware vCenter Server	VxRail embedded vCenter Server instance.
Remote site	Backup proxy server	This is a VM that is installed on the VxRail cluster.
	Backup repository	This is SAN storage or a local storage server.
	VxRail cluster	VxRail Hybrid nodes.
	VMware vCenter Server	VxRail embedded vCenter Server instance.

Table 10.5 – A summary of each component in Figure 10.12

After going through the preceding two scenarios in *Figures 10.11* and *10.12*, you got an overview of VxRail with VBR. VxRail can easily be integrated with VBR, which can deliver data protection and DR features on the local site or a remote site. The next section will discuss the key benefits of VxRail with VBR.

> **Important note**
> If you enabled a vSphere Replication session on a VxRail cluster in a local site or two sites, it is not recommended to enable Veeam VM replication on the VxRail cluster because both can deliver VM replication based on the **Recovery Point Objective (RPO)** requirements in a local site or two sites.

Benefits of VxRail with VBR

When you integrate VxRail with VBR, you enjoy all the key benefits of VBR in the VxRail cluster. VBR v11 can deliver the following key features:

- CDP
- Hardened Linux repository
- Archive tier
- Instant recovery for SQL, Oracle, and NAS
- Veeam-powered **Backup as a Service (BaaS)** and **DR as a Service (DRaaS)**

CDP

CDP is a new feature that is available on VBR v11. It is used to protect critical workloads with near-zero RPOs. It can minimize data loss and recover data with the latest state or **point in time (PIT)**. It supports short-term retention (crash-consistent), long-term retention (application-aware or crash-consistent), and permanent failover and failback. Veeam CDP is similar to Veeam VM replication (Replica). *Table 10.6* shows a feature comparison of Veeam CDP and Replica:

Features	CDP	Replica
Failover and failback	Supported	Supported
Planned failover	Not supported	Supported
Failover plan	Supported	Supported
SureReplica	Not supported	Supported
File-level recovery	Not supported	Supported
Network throttling	Supported	Supported
Preferred networks	Supported	Supported
Automation	Supported	Supported
CloudConnect	Not supported	Supported

Table 10.6 – A feature comparison of Veeam CDP and Replica

Hardened Linux repository

This feature is used to keep your data safe with malware-safe storage repositories for your backups to prevent malicious encryption and deletion. You can keep your backups and copies safe for the designated retention period.

> **Important note**
> A hardened repository is only supported on the Linux platform; it is not supported on the Microsoft Windows platform.

Archive tier

This feature is used to reduce the cost of long-term archives, replace manual tape management, and achieve **end-to-end** (**E2E**) backup life cycle management.

Instant recovery for SQL, Oracle, and NAS

You can easily recover the workloads of failed SQL and Oracle databases from Instant VM Recovery in the production environment. You can access the data directly by instant recovery of NAS, then copy the data into the production environment. *Table 10.7* shows a comparison of instant recovery between SQL and Oracle:

Instant Recovery	SQL Database	Oracle Database
Cluster configuration	Supported	Not supported
Application-aware	Supported	Supported
Archive tier	Not supported	Not supported
Automatic storage management	N/A	Supported

Table 10.7 – A comparison of instant recovery between SQL and Oracle

Veeam-powered BaaS and DRaaS

VBR v11 provides a powerful feature that can be integrated with Veeam Service Provider Console v5, and it supports the following features:

- It can monitor cloud-active workload backups that run in AWS, Azure, and GCP.
- It supports remote backup management for servers and endpoints.
- It saves storage capacity when you send off-site backup to the cloud via Cloud Connect.

After going through this section, you should now understand the key features of VBR v11. The next section will discuss the design of VxRail with VBR.

Design of VxRail with VBR

This section discusses two design scenarios of VxRail with VBR—that is, data protection of VMs in a local VxRail cluster and two VxRail clusters across two sites.

VxRail with VBR in a site

If you plan to design the data protection and recovery of VMs in a local VxRail cluster, you need to consider and collect the following information:

- What are your RPO and **Recovery Time Objective (RTO)** requirements for the data backup and recovery of VMs in the VxRail cluster? Both factors can affect your choice of backup mode for the backup policy of VMs—for example, network mode or SAN storage mode.

- What are the daily change rate and the data retention polices?

- How many protected VMs are required for backup recovery in the VxRail cluster? For example, if there are 75 VMs that need to be protected by VBR, the VBR server requires 75 instance licenses for Veeam Availability Suite Universal Subscription (VUL).

- For the software compatibility of VxRail (VMware vSphere) and VBR, you can refer to this **Knowledge Base (KB)** article: `https://www.veeam.com/kb2443`.

- The VBR server must be deployed outside of the VxRail cluster; the deployment of the VM is the recommended configuration.

- The Veeam backup repository is only available on Windows or Linux platforms.

- What is the total usable storage of protected VMs? This storage is the repository capacity for the backup repository server.

- Embedded vCenter Server and external vCenter Server instances are supported on VBR.

- When you deploy the VBR server, Microsoft SQL Server is one of the requirements. If you choose Microsoft SQL Server Express Edition, this can only be used in a small-scale production environment. Microsoft SQL Server Express Edition is limited by a database size of up to 10 GB. If you require a large-sized Microsoft SQL Server instance, you need to choose the standard edition of Microsoft SQL Server.

- The network bandwidth is recommended to be 1 GB or above if you enable Veeam VM replication across two sites.

- You can enable Veeam WAN Accelerator, which optimizes data transfer to a remote site if the network bandwidth between two sites is low.

Figure 10.13 shows a sample design of VxRail with VBR on a site:

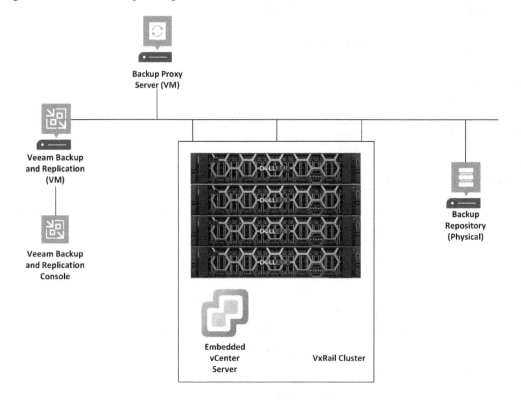

Figure 10.13 – A sample design of VxRail with VBR on a site

In this configuration, a standard VxRail stretched cluster with four nodes is used. *Table 10.8* shows the hardware configuration of each VxRail node. Each P670F node installed one 800 GB **solid-state drive** (**SSD**) (Cache tier), four 7.68 TB SSDs (Capacity tier), and two quad-port 10 GB Ethernet adapters:

VxRail model	VxRail P670F
CPU model	2 x Intel Xeon Gold 5317 3G, 12C/24T
Memory	512 GB (8 x 64 GB)
Network adapter 1	Intel Ethernet X710 Quad Port 10GbE SFP+, OCP NIC 3
Network adapter 2	Intel X710 Dual Port 10GbE SFP+ Adapter, PCIe Low Profile cards
Cache drive	1 x 800 GB SSD SAS drive
Capacity drive	4 x 7.68 TB SSD SAS drives

Table 10.8 – A sample configuration of VxRail P670F

Each VxRail P670F and VMware component contains the following software:

VxRail software release	7.0.380 build 27609715
VMware ESXi edition	7.0 Update 3d build 19482537
VMware vCenter Server	7.0 Update 3d build 19480866
VMware vSAN	7.0 Update 3d build 19482537

Table 10.9 – VxRail software releases

Table 10.10 shows the software editions of VBR:

Components	Physical Machines/VMs	Hardware and Software Editions
VBR server	VM	VBR v11a
VBR console	N/A	Bundled with VBR v11a
Backup proxy server	VM	Bundled with VBR v11a
Backup repository	Physical machine	Bundled with VBR v11a
VxRail cluster	Physical machine, VxRail cluster with four P670F nodes	VxRail software 7.0.380
VMware vCenter Server	VM, embedded vCenter Server instance	VMware vCenter Server virtual appliance 7.0 U3d build 19480866

Table 10.10 – Software editions of VBR

Each VxRail P670F installed one 800 GB SSD and four 7.68 TB SSDs; you can create a vSAN disk group with one 800 GB SSD as the cache tier and four 7.68TB SSDs as the capacity tier. Since P670F is an All-Flash model, it can support RAID-1, RAID-5, and RAID-6 site protection. Now, we will discuss the network settings of each VxRail P670F.

Each VxRail P670F has six 10 GB network ports, as shown in *Figure 10.14*. Ports **P1** and **P2** are used for the ESXi and VxRail management networks and VM networks, respectively. Ports **P3** and **P4** are used for the vSAN and vMotion networks. Ports **P5** and **P6** are used as backup networks for VMs:

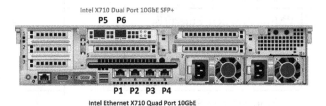

Figure 10.14 – Rear view of VxRail P670F

For the network design of the VxRail cluster, you can refer to *Table 10.11*, which shows the network layout used for a VxRail cluster:

Network Traffic	NIOC Shares	VMkernel Ports	VLAN	P1	P2	P3	P4
Management network	40%	vmk2	101	Standby	Active	Unused	Unused
vCenter Server management network	N/A	N/A	101	Standby	Active	Unused	Unused
VxRail management network	N/A	vmk1	101	Standby	Active	Unused	Unused
vSAN network	100%	vmk3	200	Unused	Unused	Active	Standby
vMotion network	50%	vmk4	100	Unused	Unused	Standby	Active
VMs	60%	N/A	N/A	Active	Standby	Unused	Unused

Table 10.11 – The network layout of VxRail

For the network design of the VBR server, you can refer to *Table 10.12*, which shows the network layout used for a VBR server:

Network Traffic	NIOC Shares	VMkernel Ports	VLAN	P5	P6
Backup network for VMs	100%	N/A	101	Active	Standby

Table 10.12 – The network layout of a VBR server

When you deploy VxRail with VBR, certain licenses are required. *Table 10.13* shows a summary of all bundled software licenses on VxRail P670F and Veeam:

Software Name	License Edition	Quantity	Remark
VMware vSphere	VMware vSphere Enterprise Plus per CPU	4	N/A
VMware vSAN	VMware vSAN Enterprise per CPU	4	You can choose vSAN Advanced or above
VMware vCenter Server	VMware vCenter Server Standard instance	1	N/A

Veeam Availability Suite Universal Subscription	Veeam Availability Suite Universal Subscription License. Includes Enterprise Plus Edition features. 30-instance pack.	30	Assume there are 30 VMs running on this VxRail cluster

Table 10.13 – A summary of all bundled software licenses on VxRail P670F

With this, you have learned the low-level design of VxRail with VBR, as illustrated in *Figure 10.13*. The next section will discuss the design of VxRail with VBR across two sites.

VxRail with VBR across two sites

Figure 10.15 shows a sample design of VxRail with VBR across two sites:

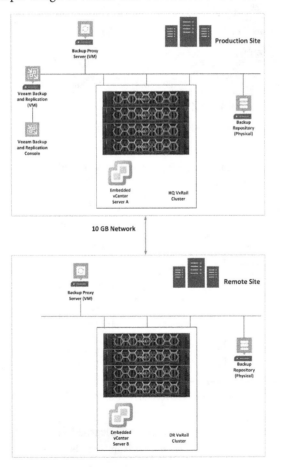

Figure 10.15 – A sample design of VxRail with VBR across two sites

In this configuration, it is a standard VxRail cluster with four nodes installed in **production** and **remote** sites. Each P670F installed one 800 GB SSD (Cache tier), four 7.68 TB SSDs (Capacity tier), one quad-port 10 GB Ethernet adapter, and two PCIe dual-port 10 GB Ethernet adapters. For VxRail software and network requirements, you can reference the low-level design of VxRail with VBR on a site shown in *Figure 10.13*.

For the network settings of each VxRail P670F, each VxRail has eight 10 GB network ports, as shown in *Figure 10.16*. Ports **P1** and **P2** are used for the ESXi and VxRail management networks and VM networks, respectively. Ports **P3** and **P4** are used for the vSAN and vMotion networks. Ports **P5** and **P6** are used as the backup networks for VMs, and ports **P7** and **P8** are used as the replication network for VMs:

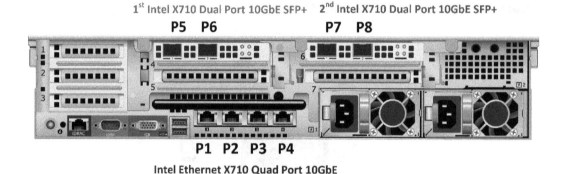

Figure 10.16 – Rear view of VxRail P670F

Table 10.14 shows the network layout used for a VBR network:

Network Traffic	NIOC Shares	VMkernel Ports	VLAN	P5	P6	P7	P8
Backup network for VMs	100%	N/A	101	Active	Standby	Unused	Unused
Replication network for VMs	100%	N/A	400	Unused	Unused	Active	Standby

Table 10.14 – The network layout of the VBR network

After going through the preceding two scenarios, you have learned the low-level design of VxRail with VBR, shown in *Figures 10.13* and *10.15*. The next section will discuss some data recovery scenarios of VxRail with VBR.

Recovery scenarios of VxRail with VBR

This last section will discuss some data recovery scenarios of VxRail with VBR, including VM recovery, faulty vCenter Server, faulty backup repository VM replication, and instant recovery.

Scenario 1

Figure 10.17 shows a sample architecture of VxRail with VBR in a single site. VBR protects the VM, and the backup images of this VM are stored in the backup repository. If the VM is corrupted in the VxRail cluster, how do we recover the data of this VM? You can restore the VM backup images based on its RPO settings from the backup repository with VBR:

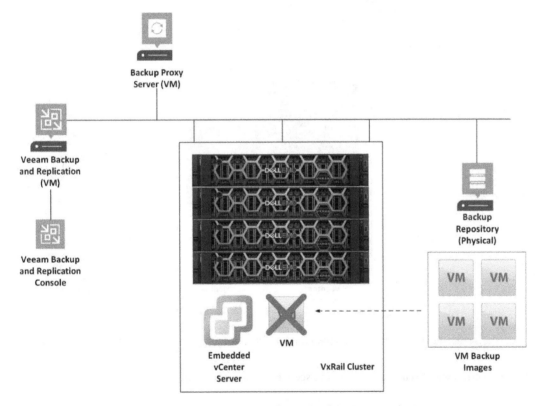

Figure 10.17 – The architecture of VxRail with VBR in a single site

The next section will discuss another failure scenario, shown in *Figure 10.18*.

Scenario 2

Figure 10.18 shows a sample architecture of VxRail with VBR in a single site. What is the impact on VBR if the embedded vCenter Server instance is unavailable in the VxRail cluster? All backup jobs will be stopped when the vCenter Server instance is not available. It is the best practice for the VMware vCenter Server to be protected by **vSphere Availability** in case the vCenter Server instance is unavailable:

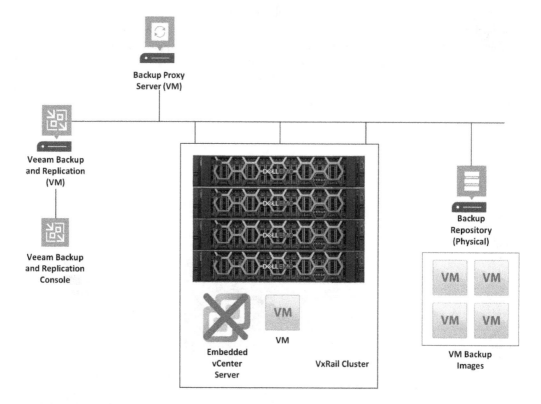

Figure 10.18 – The architecture of VxRail with VBR in a single site

The next section will discuss another failure scenario, shown in *Figure 10.19*.

Scenario 3

Figure 10.19 shows a sample architecture of VxRail with VBR across two sites. In this configuration, the VM runs on an HQ VxRail cluster, and the VM is replicated to the DR VxRail cluster by Veeam VM replication. If the VM is corrupted in the VxRail cluster, how do we recover from this VM? You can execute a Veeam failover plan to recover the VM (replica) in the DR VxRail cluster:

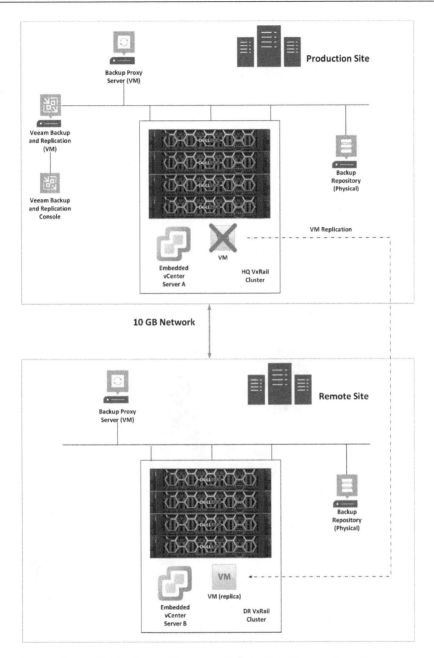

Figure 10.19 – The architecture of VxRail with VBR across two sites

The next section will discuss another failure scenario, shown in *Figure 10.20*.

Scenario 4

Figure 10.20 shows a sample architecture of VxRail with VBR in a single site. VBR protects the VM, and the backup images of this VM are stored in backup repositories **A** and **B**. One copy of the data is stored in backup repository **A**, while another copy of the data is stored in backup repository **B**. If the VM and backup repository **A** are corrupted at the same time, how do we recover from this VM? You can also recover the VM in the VxRail cluster from backup repository **B** because it is the second copy of the VM:

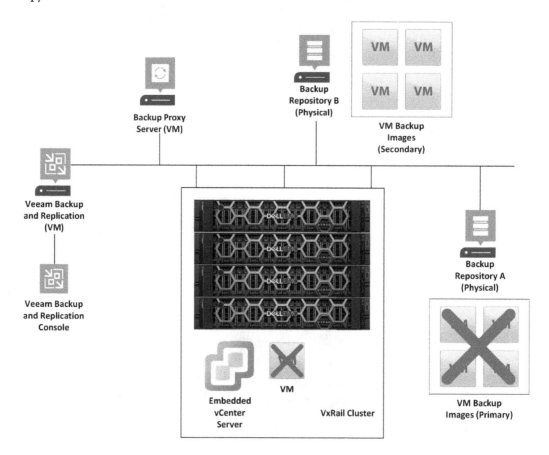

Figure 10.20 – The architecture of VxRail with VBR in a single site

The next section will discuss another failure scenario, shown in *Figure 10.21*.

Scenario 5

Figure 10.21 shows a sample architecture of VxRail with VBR in a single site. VBR protects the SQL VM, and the backup images of this VM are stored in backup repository **A**. If one of the databases is corrupted in this SQL VM, what is the fastest method of recovering this corrupted database? You can execute instant recovery of this SQL VM and restore the corrupted database into this SQL VM:

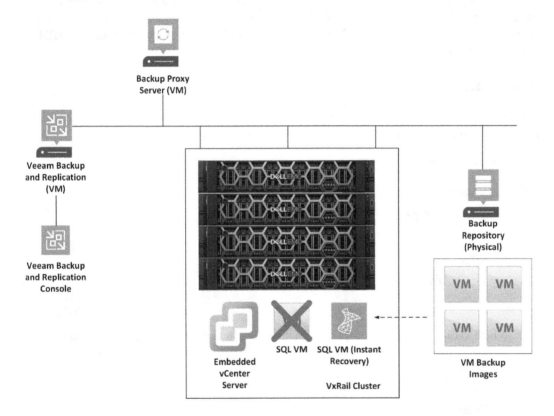

Figure 10.21 – The architecture of VxRail with VBR in a single site

After going through the preceding scenarios, you should understand the expected results on VxRail with VBR.

Summary

In this last chapter, we covered an overview and the design of VxRail with VBR, including the network, hardware, and software requirements, and some failure scenarios. VBR is a good solution for data protection and recovery and is fully integrated with Dell VxRail systems.

After going through all the chapters of this book, you should now have a good overall impression and understand the architecture of Dell VxRail 7 systems. You have also learned about the design and best practices of cluster expansion, stretched clusters, VMware **Site Recovery Manager** (**SRM**), RP4VMs on VxRail, and VBR.

Questions

As we conclude, here is a list of questions for you to test your knowledge regarding this chapter's material. You will find the answers in the *Assessments* section of the *Appendix*:

1. What can be protected by VBR v11?

    A. VMware vSphere

    B. Microsoft Hyper-V

    C. Nutanix AHV

    D. SMB

    E. Microsoft Windows

    F. All of these

2. Which Veeam component is used to store the backup images of VMs and physical machines?

    A. Veeam backup server

    B. Veeam proxy server

    C. Veeam gateway server

    D. Veeam backup repository

    E. VBR console

    F. None of these

3. Which backup rule is used for VBR?

    A. 3-2-1-1

    B. 3-2-1-2

    C. 3-2-1-1-0

    D. 3-2-2-1-0

E.   3-1-1-1-0

F.   3-2-2-2

4.   Which backup job modes are available on VBR v11?

A.   Reverse incremental backup

B.   Forever forward incremental

C.   Incremental backup

D.   Backward incremental

E.   Forward incremental

F.   Reverse and forward incremental

5.   Which platforms are supported by Veeam Agent backup?

A.   Microsoft Windows

B.   Linux

C.   VMware vSphere

D.   Oracle Unix

E.   macOS

F.   Nutanix AHV

6.   Which of the following is the simplest deployment type of VBR?

A.   One Veeam backup server, one Veeam proxy server, and one backup repository

B.   One Veeam backup server and one backup repository

C.   One Veeam backup server, one Veeam proxy server, and two backup repositories

D.   One Veeam backup server, two Veeam proxy servers, and one backup repository

E.   One Veeam backup server containing a Veeam proxy server and backup repository

F.   One Veeam backup server containing a Veeam proxy server

7.   Which of the following is an advanced deployment for VBR?

A.   One Veeam backup server, one Veeam proxy server, and one backup repository

B.   One Veeam backup server and one backup repository

C.   One Veeam backup server, one Veeam proxy server, and two backup repositories

D.   One Veeam backup server, two Veeam proxy servers, and one backup repository

E. One Veeam backup server containing a Veeam proxy server and backup repository

F. One Veeam backup server containing a Veeam proxy server

8. What are the benefits of VxRail with VBR?

A. Continuous data protection

B. Hardened Linux repository

C. Archive tier

D. Instant recovery

E. Veeam-powered BaaS and DRaaS

F. All of these

9. Which features are not included in Veeam Availability Suite (choose two)?

A. Application support, including Microsoft SQL, Exchange, Oracle, MySQL, PostgreSQL, and SAP

B. Advanced application recovery on Veeam Explorer

C. Veeam One

D. Veeam Universal License

E. Veeam Backup for Microsoft 365

F. Veeam Disaster Recovery Orchestrator

10. Which Veeam component must be a physical machine?

A. Veeam backup server

B. Veeam proxy server

C. Veeam gateway server

D. Veeam backup repository

E. VBR console

F. None of these

11. Which Veeam component must be either a physical machine or a VM (choose two)?

    A. Veeam backup server

    B. Veeam proxy server

    C. Veeam gateway server

    D. Veeam backup repository

    E. VBR console

    F. None of these

12. Which of the following scenarios can trigger all backup jobs to be stopped?

    A.

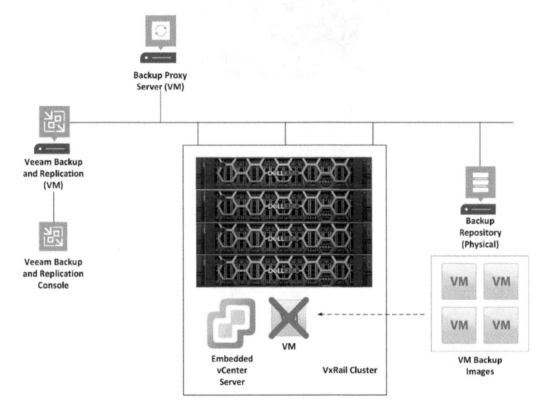

Figure 10.22 – The architecture of VxRail with VBR in a single site

B.

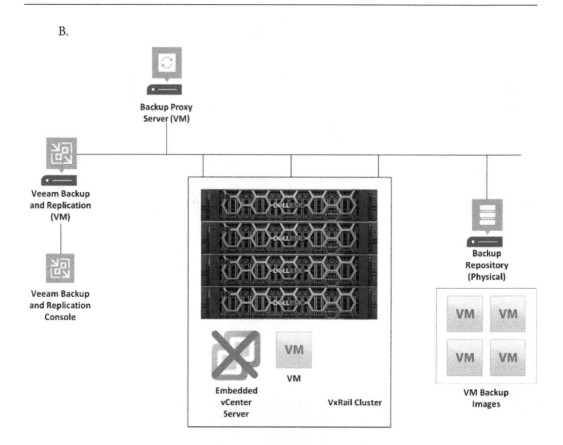

Figure 10.23 – The architecture of VxRail with VBR in a single site

C.

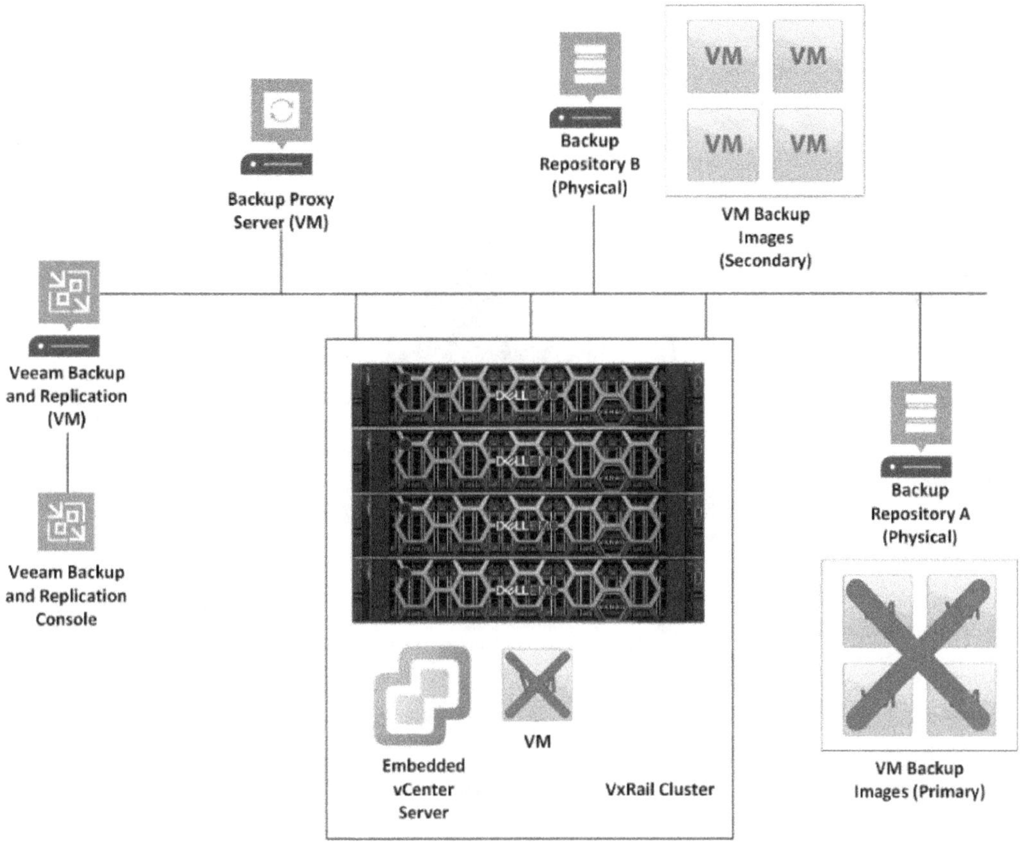

Figure 10.24 – The architecture of VxRail with VBR in a single site

D.

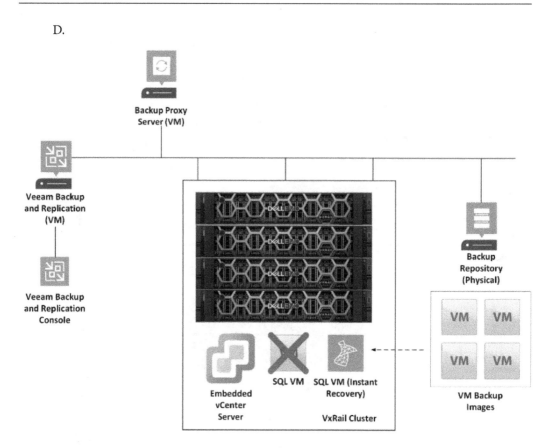

Figure 10.25 – The architecture of VxRail with VBR in a single site

Assessments

In the following pages, we will review all the practice questions from each of the chapters in this book and provide the correct answers.

Chapter 1 – Overview of VxRail Appliance 7.x System

1. E
2. D
3. A
4. C
5. D
6. B, C, D
7. E
8. E
9. B, C
10. E
11. B, E
12. A, D

Chapter 2 – Benefits of Hyper-Converged Infrastructure

1. A, C, E
2. A, C, E
3. E
4. C, D
5. D
6. A, C
7. B, C
8. E
9. C

10. B

11. D

12. B

Chapter 3 – Design of vCenter Server

1. E

2. C

3. A, B

4. D

5. A

6. D

7. A, B

8. E

9. C, D, E

10. C, D, E

11. D, E

12. D

Chapter 4 – Design of vSAN Storage Policies

1. A, E

2. G

3. C

4. E

5. D

6. F

7. B

8. C

9. E

10. C, D

11. D

12. A, B, C

Chapter 5 – Design of Cluster Expansion

1. C, E
2. B
3. B
4. D
5. D
6. D
7. B
8. A
9. A
10. D
11. A, E, F
12. B

Chapter 6 – Design of VxRail vSAN 2-Node Cluster

1. A, C, E
2. A, D
3. D
4. A, D, E
5. C, D, E
6. F
7. E
8. A
9. B
10. A, C
11. B
12. D

Chapter 7 – Design of Stretched Cluster on VxRail

1. D
2. B
3. E
4. D, E
5. B
6. C
7. B
8. C
9. A
10. D
11. A
12. D

Chapter 8 – Design of VxRail with SRM

1. B
2. A, C
3. A, C, E
4. C
5. F
6. A, C, D
7. B
8. C
9. D
10. C
11. B
12. B

Chapter 9 – Design of RecoverPoint for Virtual Machines on VxRail

1. D
2. C
3. A, C, E
4. A
5. A
6. A, E
7. A, C, D
8. D
9. A
10. B, C
11. E
12. D

Chapter 10 – Design of VxRail with Veeam Backup

1. F
2. D
3. C
4. A, B, E
5. A, B, D, E
6. E
7. A
8. F
9. E, F
10. F
11. A, E
12. B

Index

Other Books You May Enjoy

If you enjoyed this book, you may be interested in these other books by Packt:

Implementing VxRail HCI Solutions

Victor Wu

ISBN: 978-1-80107-048-5

- Set up the hardware and software requirements for a VxRail installation
- Monitor the status of VxRail appliances with the VxRail Manager plugin
- Get to grips with all the administration interfaces used to manage the VxRail appliance
- Understand vCenter roles and permissions management in the VxRail cluster
- Discover best practices for vSAN configuration in the VxRail cluster
- Find out about VxRail cluster scale-out rules and how to expand the VxRail cluster
- Deploy active-passive solutions for VxRail with VMware Site Recovery Manager (SRM)

Packt is searching for authors like you

If you're interested in becoming an author for Packt, please visit authors.packtpub.com and apply today. We have worked with thousands of developers and tech professionals, just like you, to help them share their insight with the global tech community. You can make a general application, apply for a specific hot topic that we are recruiting an author for, or submit your own idea.

Share Your Thoughts

Now you've finished *Dell VxRail System Design and Best Practices*, we'd love to hear your thoughts! Scan the QR code below to go straight to the Amazon review page for this book and share your feedback or leave a review on the site that you purchased it from.

https://packt.link/r/1804617709

Your review is important to us and the tech community and will help us make sure we're delivering excellent quality content.

Download a free PDF copy of this book

Thanks for purchasing this book!

Do you like to read on the go but are unable to carry your print books everywhere?

Is your eBook purchase not compatible with the device of your choice?

Don't worry, now with every Packt book you get a DRM-free PDF version of that book at no cost.

Read anywhere, any place, on any device. Search, copy, and paste code from your favorite technical books directly into your application.

The perks don't stop there, you can get exclusive access to discounts, newsletters, and great free content in your inbox daily

Follow these simple steps to get the benefits:

1.   Scan the QR code or visit the link below

https://packt.link/free-ebook/9781804617700

2.   Submit your proof of purchase
3.   That's it! We'll send your free PDF and other benefits to your email directly

www.ingramcontent.com/pod-product-compliance
Lightning Source LLC
Chambersburg PA
CBHW080623060326
40690CB00021B/4789